SpringerBriefs in Computer Science

Series Editors

Stan Zdonik
Peng Ning
Shashi Shekhar
Jonathan Katz
Xindong Wu
Lakhmi C. Jain
David Padua
Xuemin Shen
Borko Furht
V. S. Subrahmanian
Martial Hebert
Katsushi Ikeuchi
Bruno Siciliano

For further volumes:
http://www.springer.com/series/10028

Eli Cortez · Altigran S. da Silva

Unsupervised Information Extraction by Text Segmentation

Eli Cortez
Altigran S. da Silva
Instituto de Computação
Universidade Federal do Amazonas
Manaus
Amazonas
Brazil

ISSN 2191-5768   ISSN 2191-5776 (electronic)
ISBN 978-3-319-02596-4   ISBN 978-3-319-02597-1 (eBook)
DOI 10.1007/978-3-319-02597-1
Springer Cham Heidelberg New York Dordrecht London

Library of Congress Control Number: 2013950718

© The Author(s) 2013
This work is subject to copyright. All rights are reserved by the Publisher, whether the whole or part of the material is concerned, specifically the rights of translation, reprinting, reuse of illustrations, recitation, broadcasting, reproduction on microfilms or in any other physical way, and transmission or information storage and retrieval, electronic adaptation, computer software, or by similar or dissimilar methodology now known or hereafter developed. Exempted from this legal reservation are brief excerpts in connection with reviews or scholarly analysis or material supplied specifically for the purpose of being entered and executed on a computer system, for exclusive use by the purchaser of the work. Duplication of this publication or parts thereof is permitted only under the provisions of the Copyright Law of the Publisher's location, in its current version, and permission for use must always be obtained from Springer. Permissions for use may be obtained through RightsLink at the Copyright Clearance Center. Violations are liable to prosecution under the respective Copyright Law. The use of general descriptive names, registered names, trademarks, service marks, etc. in this publication does not imply, even in the absence of a specific statement, that such names are exempt from the relevant protective laws and regulations and therefore free for general use.
While the advice and information in this book are believed to be true and accurate at the date of publication, neither the authors nor the editors nor the publisher can accept any legal responsibility for any errors or omissions that may be made. The publisher makes no warranty, express or implied, with respect to the material contained herein.

Printed on acid-free paper

Springer is part of Springer Science+Business Media (www.springer.com)

To God

Foreword

Unsupervised Information Extraction by Text Segmentation was one of the most outstanding works presented at the 33rd Congress of the Brazilian Computer Society, in July 2013. The Authors, Eli Cortez and Altigran S. da Silva, have developed their research at the Federal University of Amazonas, Brazil, working together in the development of the tools and methodologies presented in this book. They have great expertise in the subject presented here: both are Co-founders of Neemu.com, a company that provides smart e-commerce solutions that apply state-of-the-art technologies on information retrieval and machine learning to process more than 1 million daily searches for some of the major Brazilian e-commerce companies. As such, we expect that the readers of this book will find it invaluable in the implementation of recommendation systems and in the research on similar and complementary topics.

José Viterbo
Director of Publications of the Brazilian Computer Society

Preface

Information Extraction (IE) refers to the automatic extraction of structured information such as entities, relationships between entities, and attributes describing entities from noisy unstructured textual sources. It derives from the necessity of having unstructured data stored in structured formats (tables, XML), so that it can be further queried, processed, and analyzed.

The IE problem encompasses many distinct sub-problems such as Named Entity Recognition (NER), Open Information Extraction, Relationship Extraction, and Text Segmentation. Information Extraction by Text Segmentation (IETS) is the problem of segmenting unstructured textual inputs to extract implicit data values contained in them.

In this book, we present a novel unsupervised approach for the problem of IETS. This approach relies on information available on pre-existing data to learn how to associate segments in the input string with attributes of a given domain relying on a very effective set of content-based features. The effectiveness of the content-based features is also exploited to directly learn from test data structure-based features, with no previous human-driven training, a feature unique to our approach.

Based on this approach, a number of results were obtained to address the IETS problem in an unsupervised fashion. In particular, distinct IETS methods, namely *ONDUX*, *JUDIE*, and *iForm* were implemented, evaluated, and developed.

ONDUX (On Demand Unsupervised Information Extraction) is an unsupervised probabilistic approach for IETS that relies on content-based features to bootstrap the learning of structure-based features. Structure-based features are exploited to disambiguate the extraction of certain attributes through a reinforcement step, which relies on sequencing and positioning of attribute values directly learned *on-demand* from the input texts.

JUDIE (Joint Unsupervised Structure Discovery and Information Extraction) aims at automatically extracting several semi-structured data records in the form of continuous text and having no explicit delimiters between them. In comparison with other IETS methods, including *ONDUX*, *JUDIE* faces a task considerably harder, that is, extracting information while simultaneously uncovering the underlying structure of the implicit records containing it. In spite of that, it achieves results comparable to the state-of-the-art methods.

iForm applies our approach to the task of Web form filling. It aims at extracting segments from a data-rich text given as input and associating these segments with fields from a target Web form. The extraction process relies on content-based features learned from data that was previously submitted to the Web form.

All of these methods were evaluated considering different experimental datasets, which we use to perform a large set of experiments in order to validate the presented approach and methods. These experiments indicate that the proposed approach yields high quality results when compared to state-of-the-art approaches and that it is able to properly support IETS methods in a number of real applications.

Organization

This book consists of seven chapters. Chapter 1 provides an introduction to information extraction (including how information extraction fits into the broader topics of data management), as well as a short description of the main contributions of this book. Chapter 2 then gives an overview of the existing literature and discusses related work.

The core of the book is made up by Chaps. 3–6. Chapter 3 presents the basic concepts and describes an unsupervised approach to exploit pre-existing datasets to support IETS methods. Chapter 4 presents the method called *ONDUX* and all experiments performed to evaluate its performance in comparison to other information extraction methods.

Chapter 5 presents *JUDIE*, an IE method that is able to extract information from text and capable of detecting the structure of each individual records being extracted without any user assistance. Chapter 6 presents *iForm*, a method for dealing with the Web form filling problem that relies on the presented unsupervised proposed. Finally, Chap. 7 presents the conclusions and discusses the future work.

Intended Audience

The aim of this book is to be accessible to researchers, graduate and research students, and to practitioners who work with IE and related areas. It is assumed the reader has some expertise in algorithms and data structures, and web technologies.

This book provides the reader with a broad range of IE concepts and techniques, specifically touching on all aspects of the unsupervised approach presented here. Thus, this book can help researches not only from the IE area, but also from related fields (such as databases, information retrieval, data mining, artificial intelligence, machine learning), as well as students who are interested to enter this field of research, to become familiar with the recent research developments and identify open research challenges.

This book can help practitioners to better understand the current state-of-the-art in unsupervised IE techniques. Given that in many applications domains it is not feasible to simply use or implement an existing off-the-shelf IE system without substantial adaption, it is crucial for practitioners to understand different aspects of the existing extraction methods. The technical level of this book also makes it accessible to students taking advanced undergraduate and graduate level courses on Web data management.

Acknowledgments

The Authors would like to express their gratitude to colleagues at the UFAM BDRI Group and at the InWeb Project, especially to those we collaborate in many parts of this work: Edleno Moura, Guilherme Toda, Felipe Mesquita, Alberto H. F. Laender, and Marcos Andr'e Gonçalves. We are also grateful to our families and friends for their continuous support.

This work was partially founded by projects DOMAR (CNPq 476798/2011-6), TTDSW (PRONEM/FAPEAM/CNPq), INWeb (MCT/CNPq 57.3871/2008-6), by UOL Bolsa Pesquisa program (grant 20090213165000), and by the authors' individual grants and scholarships from CNPq, CAPES and SUFRAMA.

Manaus, August 2013   Eli Cortez
                                                                         Altigran S. da Silva

Contents

Chapter 1
Introduction

Abstract The Information Extraction problem (IE) refers to the automatic extraction of structured information from noisy unstructured textual sources. This problem is a research topic in different Computer Science communities, such as: Databases, Information Retrieval, and Artificial Intelligence. This chapter provides an introduction of this problem and also an overview of how information extraction fits into the broader topics of data management. It also provides a list of the main contribution that can be found in this book.

Keywords World Wide Web · Information extraction · Textual sources · Text segmentation · Databases · Data management

1.1 Information Extraction

Over the past years, there has been a steady increase in the number and types of source of textual information available in the World Wide Web. Examples of such sources are e-shops, digital libraries, social networks, blogs, etc. In most cases, these sources are freely accessible, cover a variety of topics and subjects, provide information in distinct formats and styles, and do not impose any rigid publication format. In addition, they are constantly kept up-to-date by users and organizations. In fact, textual Web sources are typically user-oriented, i.e., they are built for users to consume their contents and the ease of interaction they provide has made the number of users that heavily interact with them grow everyday. Figure 1.1 illustrates some popular sources of textual information currently available on the Web.

Through the eyes of data management scientists, these sources constitute large repositories of valuable data on a variety of domains. Depending on the types of each source, one can find in them data referring to personal information, products, publications, companies, cities, weather, etc., from which it is possible to perform

E. Cortez and A. S. da Silva, *Unsupervised Information Extraction by Text Segmentation*, SpringerBriefs in Computer Science, DOI: 10.1007/978-3-319-02597-1_1,
© The Author(s) 2013

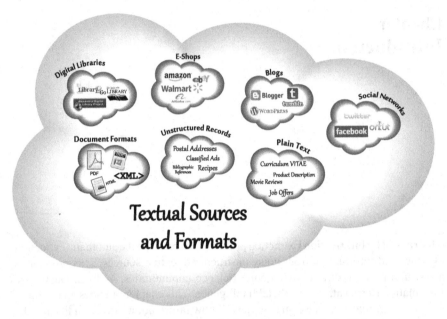

Fig. 1.1 Examples of popular sources of textual information available in the Web nowadays

such tasks as to infer relationships, to learn user preferences, and to detect trends, to name a few.

However, the same features that have made textual Web sources so useful and popular also impose important restrictions on the way the data they contain can be manipulated. Particularly, data-rich text snippets, such as product descriptions, bibliographic citations, postal addresses, classified ads, and movie reviews, are inherently unstructured and their content can hardly be subject to some form of automated processing. In addition, it is commonly difficult to automatically identify data of interest that is implicitly present in such sources embedded with pieces of nonrelevant text.

As an example, in Fig. 1.2, it is shown a real web page containing a cooking recipe. As it can be noticed, the recipe information, which constitute data-rich text snippets, as stated earlier, is embedded within web ads, free texts, cooking directions, and reviews. In addition, the data available within these snippets is loosely structured.

Nevertheless, the abundance and popularity of these online sources of relevant data have attracted a number of research efforts to address problems related to them, such as crawling (Barbosa and Freire 2007; Vidal et al. 2006), extracting (Chang et al. 2006; Laender et al. 2002; Sarawagi 2008), querying (Halevy 2012; Sardi Mergen et al. 2010), mining (Cafarella et al. 2008; Jin et al. 2009), and others (Chang et al. 2004; Madhavan et al. 2007).

In particular, the extraction problem, commonly known as *Information Extraction* (*IE*) in the literature (Sarawagi 2008), refers to the automatic extraction of structured

Fig. 1.2 Example of a real web page containing data-rich text snippets

information, such as entities, relationships between entities and attributes, describing entities from noisy unstructured sources. It derives from the necessity of having unstructured data stored in structured formats (tables, XML), so that it can be further queried, processed, and analyzed. This problem is the main subject of this work.

The IE problem encompasses many distinct subproblems such as Named Entity Recognition (NER) (Ratinov and Roth 2009; Ritter et al. 2011), Open Information Extraction (Banko et al. 2009; Mausam et al. 2012), Relationship Extraction (Fader et al. 2011; Mesquita and Barbosa 2011), and Text Segmentation (Borkar et al. 2001; Cortez et al. 2010; Sarawagi 2008).

Information Extraction by Text Segmentation (IETS) is the problem of segmenting unstructured textual inputs to extract implicit data values contained in them. Considering the practical and theoretical importance of the IETS problem (Agichtein and Ganti 2004; Borkar et al. 2001; Cortez et al. 2010, 2011; Lafferty et al. 2001; Mansuri and Sarawagi 2006; Sarawagi 2008; Zhao et al. 2008), this book presents an unsupervised approach to address it. This approach relies on pre-existing data to provide features for a learning process, alleviating the need for manually labeled data for training. Next, it is presented in more detail the definition of the IETS problem and discussed the research challenges faced.

1.2 Information Extraction by Text Segmentation

Information Extraction by Text Segmentation (IETS) is the problem of extracting attribute values occurring in implicit semi-structured data records in the form of continuous text, such as product descriptions, bibliographic citations, postal addresses, classified ads, etc. It is an important practical problem that has been frequently addressed in the recent literature (Mansuri and Sarawagi 2006; Sarawagi 2008; Zhao et al. 2008). More specifically, the main goal is to find attribute values within unstructured textual snippets. The final output of the extraction process varies; but usually, it can be transformed so as to populate a database for further processing and analysis.

To better illustrate this problem, consider Fig. 1.3. Figure 1.3a depicts a real unstructured record (postal address). This record contains relevant information, such as: person name, street name, house number, zip code, etc., and does not contain any explicit delimiter between the values composing it. Figure 1.3b shows an expected output for this example, where each segment receives a label indicating that the text segment contains a value of an attribute.

A fairly common approach to solve this problem is the use of machine learning techniques, either supervised, i.e., with human-driven training (Borkar et al. 2001; Freitag and McCallum 2000; Peng and McCallum 2006), or unsupervised, i.e., with training provided by some form of pre-existing data source (Agichtein and Ganti 2004; Chuang et al. 2007; Cortez et al. 2010, 2011; Mansuri and Sarawagi 2006; Zhao et al. 2008).

Current IETS methods, i.e., methods for solving the IETS problem, rely on probabilistic graph-based models (Lafferty et al. 2001; Sarawagi 2008) in which nodes (states) represent attributes and edges (transitions) to represent the likely structures of the data records. When properly trained, such models are able to accurately predict a sequence of labels to be assigned to a sequence of text segments corresponding to attribute values.

The learning process thus consists in capturing content-based (or state) features, which characterize the domain of the attributes (e.g., typical values, terms composing

(a) Eli Cortez - Rua 15 n 324 - Japiim 1 - 69075 - Manaus

Name	Street	Number	Neigh.	Zip	City
(b) Eli Cortez	Rua 15	n 324	Japiim 1	69075	Manaus

Fig. 1.3 Example of an unstructured textual record (**a**) and a expected output (**b**)

them, their format, etc.), and structure-based (or transition) features (e.g., the positioning and sequencing of attribute values, etc.), which characterize the structure of the records within the source text.

1.3 Main Contributions

To alleviate the need for manually labeled training data, recent IETS methods (Agichtein and Ganti 2004; Mansuri and Sarawagi 2006) rely on pre-existing datasets, such as dictionaries, knowledge bases, and references tables, from which content-based features (e.g., vocabulary, value range, writing style) can be learned. Such features are known to be very effective as state features in sequential models, such as Conditional Random Fields (CRF) (Lafferty et al. 2001). Besides saving the user effort, using pre-existing datasets also makes the process of learning content-based features less dependent from the input texts. For instance, Agichtein and Ganti (Agichtein and Ganti 2004) proposed the use of *reference tables* to learn content-based features in order to create Hidden Markov Models capable of extracting information from bibliographic references and postal addresses. Zhao et al. (2008) rely on the same idea of exploiting reference tables, but, in this case, the features are used to automatically train CRF models.

This work has further exploited this idea and has shown that content-based features learned from pre-existing datasets can also be used to bootstrap the learning of structure-based features, which are used as transition features in sequential models. Thus, it follows that these datasets allow the unsupervised learning of both content-based and structure-based features (Cortez et al. 2010, 2011).

Specifically, in this work it is presented an unsupervised approach to the IETS problem. This approach relies on information available on pre-existing data, namely *knowledge bases*, to learn how to associate segments in the input string with attributes of a given domain relying on a very effective set of content-based features. The effectiveness of the content-based features is also exploited to directly learn from test data structure-based features, with no previous human-driven training, a feature that is unique to this approach.

Based on this approach, it was produced a number of results to address the problem of information extraction by text segmentation in an unsupervised fashion. Particularly, it was developed, implemented, and evaluated distinct IETS methods.

For the case where the input unstructured records are explicitly delimited in the input text, it was proposed a method called *ONDUX* (Cortez and da Silva 2010; Cortez et al. 2010; Porto et al. 2011). *ONDUX* (On Demand Unsupervised Information Extraction) is an unsupervised probabilistic approach for IETS. Like other unsupervised IETS approaches, *ONDUX* relies on information available on pre-existing data, but, unlike previously proposed methods, it also relies on a very effective set of content-based features to bootstrap the learning of structure-based features. More specifically, structure-based features are exploited to disambiguate the extraction of certain attributes through a reinforcement step. The novel reinforcement step relies on the sequencing and positioning of attribute values directly learned *on-demand* from test data. This assigns to *ONDUX* a high degree of flexibility and considerably improves its effectiveness, as demonstrated by the experimental evaluation it is reported with textual sources from different domains, in which *ONDUX* is compared with a state-of-the-art IETS approach. Some applications use *ONDUX* to perform information extraction tasks, one example is Ciência Brasil[1] (Laender et al. 2011a,b), a research social network for Brazilian scientists.

Also a method called *JUDIE* (Cortez et al. 2011) was developed based on the unsupervised approach, for dealing with textual inputs that do not contain any explicit structural information available. *JUDIE* (Joint Unsupervised Structure Discovery and Information Extraction) is a method for automatically extracting semi-structured data records in the form of continuous text (e.g., bibliographic citations, postal addresses, classified ads, etc.) and having no explicit delimiters between them. *JUDIE* is capable of detecting the structure of each individual record being extracted without any user assistance. This is accomplished by a novel Structure Discovery algorithm that, given a sequence of labels representing attributes assigned to potential values, group these labels into individual records by looking for frequent patterns of label repetitions among the given sequence. In comparison with other IETS methods, including *ONDUX*, *JUDIE* faces a task considerably harder, that is, extracting information while simultaneously uncovering the underlying structure of the implicit records containing it. Through an extensively experimental evaluation with different datasets in distinct domains, *JUDIE* is compared with state-of-the-art information extraction methods and conclude that, even without any user intervention, it is able to achieve high quality results on the tasks of discovering the structure of the records and extracting information from them.

As it can be noticed, both *ONDUX* and *JUDIE* rely on information available on pre-existing data to perform the extraction task. To support these methods, it was presented in (Serra et al. 2011) a strategy for automatically obtaining datasets from Wikipedia. The achieved results suggest that the developed strategy is valid and effective, and that IETS methods can achieve a very good performance if the datasets generated have a reasonable number of representative values on the domain of the data to be extracted.

Finally, it is also shown how the unsupervised approach was applied by a method called *iForm* to the task of Web form filling (Toda et al. 2009, 2010). In this case, the aim is at extracting segments from a data-rich text given as input and associating these segments with fields from a target Web form. The extraction process relies on content-based features learned from data that was previously submitted to the Web form. Through extensive experimentation, it is shown that the use of the unsupervised approach in *iForm* is feasible and effective, and that it works well even when only a few previous submissions to the input interface are available, thus achieving high quality results when compared to the baseline.

---

[1] http://www.pbct.inweb.org.br/pbct/

References

Agichtein, E., & Ganti, V. (2004). Mining reference tables for automatic text segmentation. In *Proceedings of the ACM SIGKDD International Conference on Knowledge Discovery and Data Mining* (pp. 20–29), Seattle, USA.

Banko, M., Cafarella, M., Soderland, S., Broadhead, M., & Etzioni, O. (2009). *Open information extraction for the web.* PhD thesis, University of Washington.

Barbosa, L., & Freire, J. (2007). An adaptive crawler for locating hidden-web entry points. In *Proceedings of the WWW International World Wide Web Conferences* (pp. 441–450), Alberta, Canada.

Borkar, V., Deshmukh, K., & Sarawagi, S. (2001). Automatic segmentation of text into structured records. In *Proceedings of the ACM SIGMOD International Conference on Management of Data Conference* (pp. 175–186), Santa Barbara, USA.

Cafarella, M., Halevy, A., Wang, D., Wu, E., & Zhang, Y. (2008). Webtables: Exploring the power of tables on the web. *Proceedings of the VLDB Endowment, 1*(1), 538–549.

Chang, K., He, B., Li, C., Patel, M., & Zhang, Z. (2004). Structured databases on the web: Observations and implications. *ACM SIGMOD Record, 33*(3), 61–70.

Chang, C., Kayed, M., Girgis, M., & Shaalan, K. (2006). A survey of web information extraction systems. *IEEE Transactions on Knowledge and Data Engineering, 18*(10), 1411–1428.

Chuang, S., Chang, K., & Zhai, C. (2007). Context-aware wrapping: Synchronized data extraction. In *Proceedings of the VLDB International Conference on Very Large Data Bases* (pp. 699–710), Viena, Austria.

Cortez, E., & da Silva, A. S. (2010). Unsupervised strategies for information extraction by text segmentation. In *Proceedings of the SIGMOD PhD Workshop on Innovative Database Research* (pp. 49–54), Indianapolis, USA.

Cortez, E., da Silva, A., Gonçalves, M., & de Moura, E. (2010). ONDUX: On-demand unsupervised learning for information extraction. In *Proceedings of the ACM SIGMOD International Conference on Management of Data Conference* (pp. 807–818), Indianapolis, USA.

Cortez, E., da Silva, A. S., de Moura, E. S., & Laender, A. H. F. (2011). Joint unsupervised structure discovery and information extraction. In *Proceedings of the ACM SIGMOD International Conference on Management of Data Conference* (pp. 541–552), Athens, Greece.

Fader, A., Soderland, S., & Etzioni, O. (2011). Identifying relations for open information extraction. In *Proceedings of the Conference on Empirical Methods in Natural Language Processing* (pp. 1535–1545), Edinburgh, UK.

Freitag, D., & McCallum, A. (2000). Information extraction with HMM structures learned by stochastic optimization. In *Proceedings of the National Conference on Artificial Intelligence and Conference on Innovative Applications of Artificial Intelligence* (pp. 584–589), Austin, USA.

Halevy, A. (2012). Towards an ecosystem of structured data on the web. In *Proceedings of the International Conference on Extending Database Technology* (pp. 1–2), Berlin, Germany.

Jin, W., Ho, H., & Srihari, R. (2009). OpinionMiner: A novel machine learning system for web opinion mining and extraction. In *Proceedings of the ACM SIGKDD International Conference on Knowledge Discovery and Data Mining* (pp. 1195–1204), Paris, France.

Laender, A. H. F., Ribeiro-Neto, B. A., da Silva, A. S., & Teixeira, J. S. (2002). A brief survey of web data extraction tools. *SIGMOD Record, 31*(2), 84–93.

Laender, A., Moro, M., Gonçalves, M., Davis, Jr., C., da Silva, A., Silva, A., et al. (2011a). Building a research social network from an individual perspective. In *Proceedings of the International ACM/IEEE Joint Conference on Digital Libraries* (pp. 427–428), Ottawa, Canada.

Laender, A., Moro, M., Gonçalves, M., Davis Jr, C., da Silva, A., Silva, A., et al. (2011b). Ciência Brasil—the Brazilian portal of science and technology. In *Integrated Seminar of Software and Hardware (SEMISH)*, Natal, Brasil.

Lafferty, J., McCallum, A., & Pereira, F. (2001). Conditional random fields: Probabilistic models for segmenting and labeling sequence data. In *Proceedings of the ICML International Conference on Machine Learning* (pp. 282–289), Williamstown, USA.

Madhavan, J., Jeffery, S., Cohen, S., Dong, X., Ko, D., Yu, C., et al. (2007). Web-scale data integration: You can only afford to pay as you go. In *Proceedings of the CIDR Biennial Conference on Innovative Data Systems Research* (pp. 342–350), Asilomar, USA.

Mansuri, I. R., & Sarawagi, S. (2006). Integrating unstructured data into relational databases. In *Proceedings of the IEEE ICDE International Conference on Data Engineering* (pp. 29–41), Atlanta, USA.

Mausam, Schmitz, M., Soderland, S., Bart, R., & Etzioni, O. (2012). Open language learning for information extraction. In *Proceedings of the Conference on Empirical Methods in Natural Language Processing* (pp. 523–534), Jeju Island, Korea.

Mesquita, F., & Barbosa, D. (2011). Extracting meta statements from the blogosphere. In *Proceedings of the International Conference on Weblogs and Social Media*, Barcelona, Spain.

Peng, F., & McCallum, A. (2006). Information extraction from research papers using conditional random fields. *Information Processing and Management, 42*(4), 963–979.

Porto, A., Cortez, E., da Silva, A. S., & de Moura, E. S. (2011). Unsupervised information extraction with the ondux tool. In *Simpsio Brasileiro de Banco de Dados*, Florianpolis, Brasil.

Ratinov, L., & Roth, D. (2009). Design challenges and misconceptions in named entity recognition. In *Proceedings of the Conference on Computational Natural Language Learning* (pp. 147–155), Stroudsburg, USA.

Ritter, A., Clark, S., & Etzioni, O. (2011). Named entity recognition in tweets: An experimental study. In *Proceedings of the Conference on Empirical Methods in Natural Language Processing* (pp. 1524–1534), Edinburgh, UK.

Sarawagi, S. (2008). Information extraction. *Foundations and Trends in Databases, 1*(3), 261–377.

Sardi Mergen, S., Freire, J., & Heuser, C. (2010). Indexing relations on the web. In *Proceedings of the International Conference on Extending Database Technology* (pp. 430–440), Lausanne, Switzerland.

Serra, E., Cortez, E., da Silva, A., & de Moura, E. (2011). On using Wikipedia to build knowledge bases for information extraction by text segmentation. *Journal of Information and Data Management, 2*(3), 259.

Toda, G., Cortez, E., Mesquita, F., da Silva, A., Moura, E., & Neubert, M. (2009). Automatically filling form-based web interfaces with free text inputs. In *Proceedings of the WWW International World Wide Web Conferences* (pp. 1163–1164), Madrid, Spain.

Toda, G., Cortez, E., da Silva, A. S., & de Moura, E. S. (2010). A probabilistic approach for automatically filling form-based web interfaces. *Proceedings of the VLDB Endowment, 4*(3), 151–160.

Vidal, M., da Silva, A., de Moura, E., & Cavalcanti, J. (2006). Structure-driven crawler generation by example. In *Proceedings of the International ACM SIGIR Conference on Research & Development of Information Retrieval* (pp. 292–299), Seattle, USA.

Zhao, C., Mahmud, J., & Ramakrishnan, I. (2008). Exploiting structured reference data for unsupervised text segmentation with conditional random fields. In *Proceedings of the SIAM International Conference on Data Mining* (pp. 420–431), Atlanta, USA.

Chapter 2
Related Work

Abstract In the literature, different approaches have been proposed to address the problem of extracting valuable data from the Web. In this chapter is presented an overview of such approaches. It begins by presenting a broad set of Web extraction methods and tools. Following a taxonomy previously used in the literature (Laender et al. 2002), they are divided into distinct groups according to their main approach. These groups are: *Languages for Wrapper Development*, *Wrapper Induction Methods*, *NLP-based Methods*, *Ontology-based Methods*, and *HTML-aware Methods*. Next, it is specifically presented probabilistic graph-based methods, *supervised* and *unsupervised*, and discusses their main characteristics in comparison to the unsupervised approach presented in this book.

Keywords Information extraction · Wrappers · NLP · HTML · Probabilistic methods · CRF

2.1 Web Extraction Methods and Tools

By the early 2000s, several tools and methods had been discussed in the literature for extracting valuable data from the Web. A survey on this early work is presented in Laender et al. (2002), where the authors proposed a *taxonomy* for grouping different Web extraction methods and tools based on the main approach used by each method. Here, the same taxonomy was adopted. In what follows, is described the main characteristics of the methods and tools belonging to each group.

2.1.1 Languages for Wrapper Development

One of the first initiatives for addressing the problem of extracting valuable data from the Web was the use of specialized programs able to identify data of interest and map them to some suitable format as, for instance, XML or relational tables.

E. Cortez and A. S. da Silva, *Unsupervised Information Extraction by Text Segmentation*, SpringerBriefs in Computer Science, DOI: 10.1007/978-3-319-02597-1_2, © The Author(s) 2013

These programs are called *wrappers*. Different *languages* were specially designed to assist users in developing wrappers. Such languages were proposed as alternatives to general purpose languages such as Perl and Java, which were prevalent at that time for this task.

Some of the best known tools that adopt this approach are Minerva (Crescenzi and Mecca 1998), TSIMMIS (Hammer et al. 1997), and Web-OQL (Arocena and Mendelzon 1998). Although such languages provide effective approaches for wrapper generation, their main drawback is that they required manual wrapper development. Due to such a limitation, efforts have been made to automate the wrapper generation process.

2.1.2 Wrapper Induction Methods

There were also efforts to use machine-learning techniques to semi-automatically induce wrappers (Hsu and Dung 1998; Kushmerick 2000; Muslea et al. 2001). In general, these approaches consist of using training examples to generate automata that recognize instances in contexts similar to the ones of the given examples.

The approach proposed by Kushmerick (2000) and adopted in the WEIN system relies on examples from the source to be wrapped. The main drawbacks of this work are: (1) it does not deal with missing or out-of-order components and (2) although it identifies the need for extraction of complex objects present in nested structures, the solution provided is computationally intractable and has not been implemented.

These two features of semi-structured data extraction are addressed in SoftMealy (Hsu and Dung 1998) and Stalker (Muslea et al. 2001). Both systems also generate wrappers, generalizing the given examples through machine-learning techniques, and are very effective in wrapping several types of Web pages. The main problem with SoftMealy is that every possible absence of a component and every different ordering of the components must be represented beforehand by an example. Stalker (Muslea et al. 2001) can deal with such variations in a much more flexible way since each object component is extracted independently through a top-down decomposition procedure.

The main drawback to all these approaches is that the extraction process relies on the knowledge of the structure of HTML pages. In WEIN and SoftMealy, for example, pages are assumed to have a defined structure (e.g., a head, then a body with a set of tuples, and then a tail) that must be flat. This prevents the exclusive extraction of the objects (or subobjects) of interest and might generate extraction difficulties if unwanted text portions (such as advertisements) occur between tuples or tuple components in the page body. In Stalker, the extraction of nested objects is possible but the approach also relies on a previous description of the entire source page.

2.1.3 NLP-Based Methods

Besides wrapper induction, there were other approaches for learning extraction patterns that were more suitable for extracting data from semi-structured texts such as newspaper classified advertisements, seminar announcements, and job posting, which present grammatical elements. In general, these approaches use techniques typical of Natural Language Processing (i.e., semantic class, part-of-the-speech tagging, etc.) sometimes combined with the recognition of syntactic elements (delimiters). This is the case of Rapier (Mooney 1999) and SRV (Freitag and McCallum 2000). WHISK (Soderland 1999) goes beyond and addresses a large spectrum of types of documents ranging from rigidly formatted to free text. For formatted text, this system has a behavior that is closer to wrapper induction systems like WEIN (Kushmerick 2000).

Recently, several new methods that also explore Natural Language Processing techniques have been proposed to deal with the *Open Information Extraction* (Etzioni et al. 2008) problem. In this context, the goal is to perform Web scale extraction from all types of textual documents available on the Web. The system makes a single data-driven pass over its dataset and extracts a large set of relational tuples without requiring any human input. Banko et al. (2007, 2009) introduced a system called TEXTRUNNER, an open information extraction system that is able to extract tuples from large datasets and also allow their exploration via user queries. Different from the presented unsupervised approach, these open information extraction approaches rely heavily on linguistic information requiring the presence of grammatical elements.

2.1.4 Ontology-Based Methods

An ontology-based approach to extracting data from Web sources was proposed by Embley et al. (1999a). This approach uses a semantic data model to provide an ontology that describes the data of interest, including relationships, lexical appearances, and context keywords. By parsing this ontology, a relational database schema and a constant/keyword recognizer are automatically generated, which are then used to extract the data that will populate the database. Prior to the application of the ontology, the approach requires the application of an automatic procedure to extract chunks of text containing data "items" (or records) of interest (Embley et al. 1999b). Then, the extraction process proceeds from the set of records extracted. Not only does this approach require the user to provide a conceptual description of the data to be extracted, but relies mainly on the expected contents of the pages, which is anticipated by the prespecified ontology. Further, this approach requires a specialist to build the ontology using a notation specially designed for this task.

2.1.5 HTML-Aware Methods

Crescenzi et al. (2001) proposed RoadRunner, a method that heavily explores the inherent features of HTML documents to automatically generate wrappers. Road-Runner works by comparing the HTML structure of two (or more) given sample pages belonging to a same "page class," generating as a result a schema for the data contained in the pages. To accurately capture all possible structural variations occurring on pages of a same page class, it is possible to provide more than two sample pages. The extraction process is based on an algorithm that compares the tag structure of the sample pages and generates regular expressions that handle structural mismatches found between the two structures. It should be noted that the process is fully automatic and no user intervention is required, a feature that was unique to RoadRunner by that time. Although very effective, RoadRunner relies on specific HTML features to uncover the structure of the objects to be extracted. In such cases, fully automated tools tend to make a lot of misinterpretations, in the sense that they can extract several unwanted data.

There are also methods that rely on the representation of the HTML documents as *DOM* trees. Reis et al. (2004) and Dalvi et al. (2009) propose techniques based on tree edit distance to perform the extraction task. In Zhao et al. (2005) the authors propose the use of both the visual content of the HTML pages as displayed on a browser and the HTML DOM tree to perform the extraction.

More recently, a set of methods has been proposed for detecting and extracting information available on HTML tables. A system that is able to explore tabular information available within HTML pages is described by Cafarella et al. (2008). For this, the *Webtables* system relies on the HTML markup to automatically detect the occurrence of tables and then extract attribute-value pairs. Following the same idea of exploring HTML structures, such as tables and lists, Elmeleegy et al. (2009) propose a technique that is able to not only extract information from HTML tables, but also lists, thus combining HTML markup characteristics with string alignment.

As it can be noticed, all of these approaches rely on the regularity of HTML documents and depend heavily on the HTML tags (document structure) to extract information of interest. In some cases, this assigned to these approaches good extraction results, however, precludes their usage in a large number of textual sources that are available on the Web. As seen in Fig. 1.1, the scenario that information extraction approaches faces nowadays includes textual sources in different formats and styles, and more specifically, free texts without any tag to explicitly indicate data of interest. In order to deal with these general textual sources the use of probabilistic graph-based approaches has been proposed, as described below.

2.2 Probabilistic Graph-Based Methods

Due to limitations of the extraction methods that are based on the HTML structure of Web pages, new methods, based on probabilistic graph-based approaches such as Hidden Markov Models (HMM) and Conditional Random Fields (CRF), were created to tackle the problem of extracting valuable data from textual sources. A fairly common approach to solve this problem is the use of machine-learning techniques, either supervised, i.e., with human-driven training, or unsupervised, i.e., with training provided by some form of pre-existing data source.

2.2.1 Supervised Probabilistic Graph-Based Methods

One of the first approaches in the literature addressing the extraction problem with a probabilistic graph-based approach was proposed by Freitag and McCallum (2000). It consisted in generating independent Hidden Markov Models (HMM) for recognizing values of each attribute. This approach was extended in the DATAMOLD tool (Borkar et al. 2001), in which attribute-driven (or *internal*) HMM are nested as states of *external* HMM. These external HMM aim at modeling the sequencing of attribute values on the implicit records. Internal and external HMM are manually trained with user-labeled text segments. Experiments over two real-life datasets yielded very good results in terms of the accuracy of the extraction process.

Later, *Conditional Random Fields (CRF)* models were proposed as an alternative to HMM for the extraction of valuable information from text (Lafferty et al. 2001). In comparison with HMM, CRF models are suitable for modeling problems in which state transitions and emissions probabilities may vary across hidden states, depending on the input sequence. Peng and McCallum (2006) proposed a supervised method for extracting bibliographic data from research papers based on CRF that showed good results in the experimental evaluation they conducted.

Kristjansson et al. (2004) also proposed the use of CRF to the task of filling Web forms with values available in unstructured texts. In this context, it is needed to extract valuable data from these texts and submit them to a predefined Web form with different form fields. Their interactive information extraction system assists the user in filling in form fields while giving the user confidence in the integrity of the data. The user is presented with an interactive interface that allows both the rapid verification of automatic field assignments and the correction of errors.

Although effective, these supervised information extraction approaches based on graphical models, such as HMM and CRF, usually require users to label a large amount of training input documents. There are cases in which training data is hard to obtain, particularly when a large number of training instances is necessary to cover several features of the test data.

2.2.2 *Unsupervised Probabilistic Graph-Based Methods*

To address the problem of requiring large amounts of manually created training sets, recent approaches presented in the literature propose the use of pre-existing data for easing the training process (Agichtein and Ganti 2004; Cortez et al. 2007; Mansuri and Sarawagi 2006; Zhao et al. 2008). These approaches take advantage of the existence of large amounts of structured datasets that can be used with little or no user effort.

According to the strategy of relying on pre-existing data, models for recognizing values of an attribute are generated from values of this attribute occurring in a dataset previously available. Mansuri and Sarawagi (2006) proposed a method based on Conditional Random Fields to extract valuable data from unstructured textual portions. The proposed method relies on pre-existing data to learn content-based features and hand-labeled training sets to learn structure-related features.

Agichtein and Ganti (2004) and Zhao et al. (2008) proposed methods that are able to train a model relying only on a pre-existing dataset and, then, use it for recognizing values of attributes among segments of the input string. No manually labeled training input strings are required for this. Once attribute values are recognized, records can be extracted. These methods assume that attributes values in the input text follow a single global order, which is learned from a sample batch of the test instances. The difference between the methods proposed by Agichtein and Ganti and the one proposed by Zhao et al. is that the first relies on Hidden Markov Models and the second relies on Conditional Random Fields. Despite this, both follow the same assumptions regarding a global attribute order in the input text.

The main difference between the presented approach and the ones presented by Agichtein and Ganti, Mansuri and Sarawagi, and Zhao et al., is the way that structure-related features (Sarawagi 2008) are learned. In the presented unsupervised approach these features, when necessary, are captured by a specific model, which, as demonstrated in the experiments, is flexible enough to assimilate and represent variations in the order of attributes in the input texts and can be learned without user-provided training. The methods proposed by Agichtein and Ganti (2004) and Zhao et al. (2008) are also capable of automatically learning structure-related features, but they cannot handle distinct orderings on the input, since they assume a single total order for the input texts. These make the application of these methods difficult to a range of practical situations. Thus, in practical applications, the presented unsupervised approach can be seen as the best alternative. The method proposed in Mansuri and Sarawagi (2006) can handle distinct ordering, but user-provided training is needed to learn the structure-related features, similar to what happens with the standard supervised CRF model, thus increasing the user dependency and the cost to apply the method in several practical situations.

A similar strategy is used by Chuang et al. (2007). However, when extracting data from a source in a given domain, this approach may take advantage not only from pre-existing datasets, but also from other sources containing data on the same domain, which is extracted simultaneously from all sources using a two-state HMM for each attribute. Record extraction is addressed in an unsupervised way by aligning records from the sources being extracted.

FLUX-CiM (Cortez et al. 2007, 2009) is an unsupervised approach for extracting metadata from bibliographic citations that rely on the same ideas adopted by the unsupervised approach presented here. While FLUX-CiM also relies on content-based features learned from pre-existing data, it uses a set of domain-specific heuristics based on assumptions regarding bibliographic metadata to perform the extraction task. This includes the use of punctuation as attribute value delimiters, the occurrence of single values for attributes other than author names, etc. Thus, the presented unsupervised approach can be seen as a generalization of FLUX-CiM.

Michelson and Knoblock (2007) presented an unsupervised approach to exploit pre-existing data for extraction. To accomplish this, initially the user has to specify a large repository with distinct sets of pre-existing data. Once this repository is chosen, using simple vector-space model similarities between the input text and the available sets of pre-existing data, the system automatically finds the most suitable set for the given extraction task. Now that a set of pre-existing data was chosen, the system relies on predefined string distance metrics such as Jaro-Winkler and Smith-Waterman, and fine-tuned thresholds to perform the extraction of valuable data. This work differs from the presented unsupervised approach in the sense that it relies on the use of predefined string similarity functions other than content-based features based on vocabulary. Moreover, the proposed system requires the availability of large pre-existing datasets in order to perform the extraction task. In the unsupervised approach presented here, this is alleviated since, when possible, it is able to automatically induce structure-related features from content-based features, helping the extraction process.

In order to support these unsupervised extraction methods that have been recently proposed in the literature, Chiang et al. (2012) developed a system called *AutoDict* that is able to automatically discover dictionaries to support unsupervised probabilistic graph-based methods. Moreover, Serra et al. (2011) show that Wikipedia can be used to support information extraction methods. Thus, these works show that it is feasible to acquire pre-existing structured datasets in order to create unsupervised extraction methods.

References

Agichtein, E., & Ganti, V. (2004). Mining reference tables for automatic text segmentation. *Proceedings of the ACM SIGKDD International Conference on Knowledge Discovery and Data Mining* (pp. 20–29). USA: Seattle.

Arocena, G., & Mendelzon, A. (1998). Weboql: Restructuring documents, databases and webs. *Proceedings of the IEEE ICDE International Conference on Data Engineering* (pp. 24–33). USA: Orlando.

Banko, M., Cafarella, M., Soderland, S., Broadhead, M., & Etzioni, O. (2009). Open information extraction for the web. PhD thesis, University of Washington, Washington.

Banko, M., Cafarella, M. J., Soderland, S., Broadhead, M., & Etzioni, O. (2007). Open information extraction from the web. *Proceedings of the IJCAI International Joint Conference on Artificial Intelligence* (pp. 2670–2676). India: Hyderabad.

Borkar, V., Deshmukh, K., & Sarawagi, S. (2001). Automatic segmentation of text into structured records. *Proceedings of the ACM SIGMOD International Conference on Management of Data Conference* (pp. 175–186). USA: Santa Barbara.

Cafarella, M., Halevy, A., Wang, D., Wu, E., & Zhang, Y. (2008). Webtables: Exploring the power of tables on the web. *Proceedings of the VLDB Endowment, 1*(1), 538–549.

Chiang, F., Andritsos, P., Zhu, E., & Miller, R. (2012). Autodict: Automated dictionary discovery. *Proceedings of the IEEE ICDE International Conference on Data Engineering* (pp. 1277–1280). USA: Washington.

Chuang, S., Chang, K., & Zhai, C. (2007). Context-aware wrapping: synchronized data extraction. *Proceedings of the VLDB International Conference on Very Large Data Bases* (pp. 699–710). Austria: Viena.

Cortez, E., da Silva, A., Gonçalves, M., Mesquita, F., & de Moura, E. (2007). FLUX-CIM: flexible unsupervised extraction of citation metadata. *Proceedings of the ACM/IEEE JCDL Joint Conference on Digital Libraries* (pp. 215–224). Canada: Vancouver.

Cortez, E., da Silva, A. S., Gonçalves, M. A., Mesquita, F., & de Moura, E. S. (2009). A flexible approach for extracting metadata from bibliographic citations. *Journal of the American Society for Information Science and Technology, 60*(6), 1144–1158.

Crescenzi, V., & Mecca, G. (1998). Grammars have exceptions. *Information Systems, 23*(8), 539–565.

Crescenzi, V., Mecca, G., & Merialdo, P. (2001). Roadrunner: Towards automatic data extraction from large web sites. *Proceedings of the VLDB International Conference on Very Large Data Bases* (pp. 109–118). Italy: Rome.

Dalvi, N., Bohannon, P., & Sha, F. (2009). Robust web extraction: an approach based on a probabilistic tree-edit model. *Proceedings of the ACM SIGMOD International Conference on Management of Data Conference* (pp. 335–348). Rhode Island, USA: Providence.

Elmeleegy, H., Madhavan, J., & Halevy, A. (2009). Harvesting relational tables from lists on the web. *Proceedings of the VLDB Endowment, 2*(1), 1078–1089.

Embley, D., Campbell, D., Jiang, Y., Liddle, S., Lonsdale, D., Ng, Y., et al. (1999a). Conceptual-model-based data extraction from multiple-record web pages. *Data and Knowledge Engineering, 31*(3), 227–251.

Embley, D., Jiang, Y., & Ng, Y. (1999b). Record-boundary discovery in web documents. *ACM SIGMOD Record, 28*(2), 467–478.

Etzioni, O., Banko, M., Soderland, S., & Weld, D. (2008). Open information extraction from the web. *Communications of the ACM, 51*(12), 68–74.

Freitag, D., & McCallum, A. (2000). Information extraction with HMM structures learned by Stochastic optimization. *Proceedings of the National Conference on Artificial Intelligence and Conference on Innovative Applications of Artificial Intelligence* (pp. 584–589). USA: Austin.

Hammer, J., McHugh, J., & Garcia-Molina, H. (1997). Semistructured data: The tsimmis experience. *Proceedings of the East-European Symposium on Advances in Databases and Information Systems* (pp. 1–8). Russia: St. Petersburg.

Hsu, C., & Dung, M. (1998). Generating finite-state transducers for semi-structured data extraction from the web. *Information systems, 23*(8), 521–538.

Kristjansson, T., Culotta, A., Viola, P., & McCallum, A. (2004). Interactive information extraction with constrained conditional random fields. *Proceedings of the AAAI National Conference on Artificial Inteligence* (pp. 412–418). San Jose: USA.

Kushmerick, N. (2000). Wrapper induction: Efficiency and expressiveness. *Artificial Intelligence, 118*(1–2), 15–68.

Laender, A. H. F., Ribeiro-Neto, B. A., da Silva, A. S., & Teixeira, J. S. (2002). A brief survey of web data extraction tools. *SIGMOD Record, 31*(2), 84–93.

Lafferty, J., McCallum, A., & Pereira, F. (2001). Conditional random fields: Probabilistic models for segmenting and labeling sequence data. *Proceedings of the ICML International Conference on Machine Learning* (pp. 282–289). USA: Williamstown.

Mansuri, I. R., & Sarawagi, S. (2006). Integrating unstructured data into relational databases. *Proceedings of the IEEE ICDE International Conference on Data Engineering* (pp. 29–41). USA: Atlanta.

Michelson, M., & Knoblock, C. (2007). Unsupervised information extraction from unstructured, ungrammatical data sources on the world wide web. *International Journal on Document Analysis and Recognition, 10*(3), 211–226.

Mooney, R. (1999). Relational learning of pattern-match rules for information extraction. *Proceedings of the National Conference on Artificial Intelligence* (pp. 328–334). USA: Orlando.

Muslea, I., Minton, S., & Knoblock, C. A. (2001). Hierarchical wrapper induction for semistructured information sources. *Autonomous Agents and Multi-Agent Systems, 4*(1–2), 93–114.

Peng, F., & McCallum, A. (2006). Information extraction from research papers using conditional random fields. *Information Processing and Management, 42*(4), 963–979.

Reis, D. C., Golgher, P. B., Silva, A. S., & Laender, A. F. (2004). Automatic web news extraction using tree edit distance. *Proceedings of the WWW International World Wide Web Conferences* (pp. 502–511). USA: New York.

Sarawagi, S. (2008). Information extraction. *Foundations and Trends in Databases, 1*(3), 261–377.

Serra, E., Cortez, E., da Silva, A., & de Moura, E. (2011). On using wikipedia to build knowledge bases for information extraction by text segmentation. *Journal of Information and Data Management, 2*(3), 259.

Soderland, S. (1999). Learning information extraction rules for semi-structured and free text. *Machine learning, 34*(1), 233–272.

Zhao, C., Mahmud, J., & Ramakrishnan, I. (2008). Exploiting structured reference data for unsupervised text segmentation with conditional random fields. *Proceedings of the SIAM International Conference on Data Mining* (pp. 420–431). USA: Atlanta.

Zhao, H., Meng, W., Wu, Z., Raghavan, V., & Yu, C. (2005). Fully automatic wrapper generation for search engines. *Proceedings of the WWW International World Wide Web Conferences* (pp. 66–75). Japan: Chiba.

Chapter 3
Exploiting Pre-Existing Datasets to Support IETS

Abstract This chapter describes in detail a new approach for exploiting preexisting datasets to support Information Extraction by Text Segmentation methods. First, it presents a brief overview of the approach and introduces the concept of knowledge base. Next, it discusses all the steps involved in the unsupervised approach, including how to learn content-based features from knowledge bases, how to automatically induce structure-based features with no previous human-driven training, a feature that is unique to this approach, and how to effectively combine these features to label segments of a text input.

Keywords Information extraction · Unsupervised approach · Text segmentation · Databases · Structured data · Knowledge bases · Markov models

3.1 Overview

Consider a set of data-rich input text snippets from which we need to extract data containing in them. it is assumed that all snippets in this set belong to the same application domain (e.g., product descriptions, bibliographic citations, postal addresses, real estate classified ads, etc). It is also assumed the existence of a dataset on the same domain as the input set, which is called *Knowledge Base*.

The presented unsupervised approach to tackle the information extraction by text segmentation problem, relies on the following steps, which are illustrated in Fig. 3.1: (1) learn content-based features from a knowledge base, (2) use the learned content-based features in an initial extraction process, (3) explore the outcome of the initial extraction process to automatically induce structure-based features, and (4) combine content-based features with structure-based features to achieve a final extraction result. Thus, this approach relies on the hypothesis that the usage of knowledge bases allow for the unsupervised learning of both content-based and structure-based features.

E. Cortez and A. S. da Silva, *Unsupervised Information Extraction by Text Segmentation*,   19
SpringerBriefs in Computer Science, DOI: 10.1007/978-3-319-02597-1_3,
© The Author(s) 2013

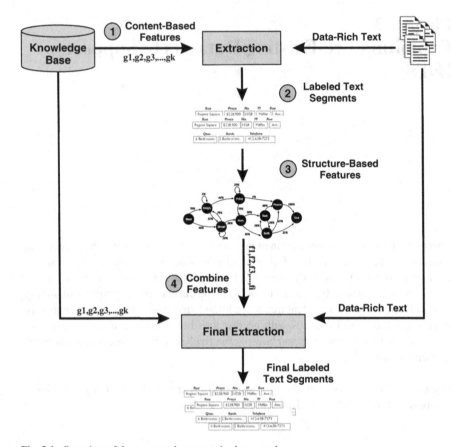

Fig. 3.1 Overview of the presented unsupervised approach

Different content-based features can be learned from the knowledge implicitly encoded in the knowledge bases, which are exploited by this approach. These features are: (1) attribute vocabulary, (2) attribute value range, and (3) attribute value format. A very important point to stress regarding these content-based features is the fact that they can be computed from previously available knowledge bases and, thus, they are independent of the target input text corpus, that is, these features is *input-independent*.

The attribute vocabulary feature exploits the common terms often shared by values of textual attributes. The attribute value range feature specifically deals with numeric attributes using the average and the standard deviation of the values of numeric attributes available on the knowledge base. Finally, The attribute value format feature exploits the writing styles often used to represent values of different attributes in the knowledge base (e.g., url, date, telephone). It is assumed that these features exploit different properties of the attribute domain, thus, it can be said they are independent, what allows us to combine them by means of the Bayesian disjunctive operator *or*, also known as *Noisy-OR-Gate* (Pearl and Shafer 1988).

As it can be noticed by the experiments that were performed, the unsupervised approach here presented is able to perform the extraction of valuable data relying only on content-based features. However, there are cases in which it can further exploit these features to automatically induce structure-based features and improve the quality of the extraction results. For computing such structure-based features, it is common to use a graph model that represents the likelihood of attribute transitions within the input text (or any other input text from the same source). It used a probabilistic HMM-like graph model that is called PSM (Positioning and Sequencing Model). With the structure-based features in hand, we can use them to improve the initial extraction that resorted only on content-based features.

In the following, is presents the concept of knowledge base and shows how to learn content-based features from such knowledge bases. It is also shown how to induce structure-based features from content features and how to automatically combine these features using a Bayesian disjunctive operator.

3.2 Knowledge Bases

A Knowledge Base is a set of pairs $K = \{\langle a_1, O_1 \rangle, \ldots, \langle a_n, O_n \rangle\}$ in which each a_i is a distinct attribute and O_i is a set of strings $\{o_{i,1}, \ldots, o_{i,n_i}\}$ called *occurrences*. Intuitively, O_i is a set of strings representing plausible or typical values for an attribute a_i.

In Fig. 3.2 it illustrated a very simple example of a knowledge base which includes only four attributes: Neighborhood, Street, Bathrooms, and Phone. Notice that, a knowledge base contains common words that usually occur as attribute values, and given the fact that there are several sources of structured information available, such as FreeBase and Wikipedia, its construction process can be regarded as simple (Chiang et al. 2012; Serra et al. 2011).

In fact, given a data source on a certain domain that includes values associated with fields or attributes, building a knowledge base is a simple process that consists in creating pairs of attributes and sets of occurrences. Notice that the knowledge bases implicitly encode *domain knowledge*. Thus, they are a very suitable source for learning content-based features.

Some IETS methods (Agichtein and Ganti 2004; Mansuri and Sarawagi 2006) rely on preexisting datasets such as dictionaries and references tables, from which

$$K = \{\langle Neigh., O_{Neigh.} \rangle, \langle Street, O_{Street} \rangle, \langle Bathrooms, O_{Bathrooms}, Phone, O_{Phone} \rangle\}$$
$$O_{Neighborhood} = \{\text{"Regent Square", "Milenight Park"}\}$$
$$O_{Street} = \{\text{"Regent St.", "Morewood Ave.", "Square Ave. Park"}\}$$
$$O_{Bathrooms} = \{\text{"Two Bathrooms", "5 Bathrooms"}\}$$
$$O_{Phone} = \{\text{"(323) 462-6252", "171 289-7527"}\}$$

Fig. 3.2 A simple example of a Knowledge Base

content-based features (e.g., vocabulary, value range, format) can be learned. For instance, Mansuri and Sarawagi (2006) proposed a method that uses words stored in dictionaries. The Unsupervised CRF method proposed in (Zhao et al. 2008) requires full records stored in reference tables. The presented methods, ONDUX (Cortez et al. 2010) and JUDIE (Cortez et al. 2011) rely on sets of *attribute values* stored on a knowledge base, as defined earlier. To simplify the terminology, it will be used, from now on, the term knowledge base to refer to all of these kinds of datasets.

3.3　Learning Content-Based Features

All content-based features used in the presented unsupervised approach can be computed from a knowledge base. Consider an attribute A and let v_A be a set of typical values for this attribute. Then, for any segment of tokens $\langle x_i, \ldots, x_j \rangle$ from the input text, it can be computed the value of a feature function $g^k(\langle x_i, \ldots, x_j \rangle, A)$. Intuitively, g^k returns a real number that measures how well a hypothetical value formed by tokens in the text segment $\langle x_i, \ldots, x_j \rangle$ follows some property of the values in the domain of A represented by v_A (Sarawagi 2008). Obviously, the accuracy of such functions often depends on how representative v_A is with respect to the values in the domain of A. The content-based features considered in the presented approach are described below.

3.3.1　Attribute Vocabulary

This feature exploits the common vocabulary often shared by values of textual attributes (e.g., neighborhood and street names, author names, recipe ingredients, etc.). To capture this property, the presented unsupervised approach resorts to a function called *AF* (Attribute Frequency) (Mesquita et al. 2007), which estimates the similarity between a given value and the set of values of an attribute. In this case, the function *AF* is used to estimate the similarity between the content of a candidate value s and the values of an attribute A represented in the knowledge base. Function *AF* is defined as follows:

$$AF(s, A) = \frac{\displaystyle\sum_{t \in T(A) \cap T(s)} fitness(t, A)}{|T(s)|} \qquad (3.1)$$

In Eq. 3.1, $T(A)$ is the set of all terms found in the values of attribute A in the knowledge base and $T(s)$ is the set of terms found in a candidate value s. The function $fitness(t, A)$ evaluates how typical a term t is among the values of attribute A. It is computed as follows:

$$fitness(t, A) = \frac{f(t, A)}{N(t)} \times \frac{f(t, A)}{f_{max}(A)} \tag{3.2}$$

where $f(t, A)$ is the number of distinct values of A that contain the term t, $f_{max}(A)$ is the highest frequency of any term among the values of A, and $N(t)$ is the total number of occurrences of the term t in all attributes represented in the knowledge base.

The first fraction in Eq. 3.2 expresses the likelihood of term t to be part of a value of A according to the knowledge base. This fraction is multiplied by a normalization factor in the second fraction. This prevents attributes with many values in the knowledge base from dominating and is also useful for making the term frequency comparable among all attributes.

As an example, consider the text segment s = "Regent Park", the knowledge base presented in Fig. 3.2 and the attribute Neighborhood available in this knowledge base. According to this setting, A = neighborhood, T(neighborhood) = $\{regent, square,$
$milenight, park\}$ and $T(s) = \{regent, park\}$.

It can be noticed that although we could have used any other similarity function, for instance, based on the Vector Space Model (Salton et al. 1975), experiments reported in the literature (Cortez et al. 2007, 2010, 2011; Mesquita et al. 2007) have shown that AF is very effective for dealing with small portions of texts such as the ones typically found in candidate values. It is also worth mentioning that the implementation of the unsupervised approach uses inverted indexes over the knowledge base to speed up the computation of this content-based feature.

3.3.2 Attribute Value Range

For the case of numeric candidate values (e.g., page number, year, phone number, price, quantity, etc.) textual similarity functions such as *AF* (Eq. 3.1) do not work properly. Thus, for dealing with these candidate values, a proper content-based feature function is needed. it is assumed, as proposed in Agrawal and Chaudhuri (2003), that the values of numeric attributes follow a Gaussian distribution. Based on this assumption, it is measured the similarity between a numeric value v_s present in a candidate value s and the set of values v_A of an attribute A in the knowledge base, by evaluating how close v_s is from the mean value of v_A according to its probability density function. For that, it is used the function *NM* (Numeric Matching) normalized by the maximum probability density of v_A, which is reached when a given value is equal to the average.[1] This function is given by

$$NM(s, A) = e^{-\frac{v_s - \mu_A}{2\sigma_A^2}} \tag{3.3}$$

---

[1] The maximum probability density of v_A is $1/\sqrt{2\pi\sigma^2}$.

where σ_A and μ_A are, respectively, the standard deviation and the average of values in v_A, and v_s is the numeric value of s. Notice that when v_s is close to the average of values in v_A, $NM(s, A)$ is close to 1. When v_s assumes values far from the average, the similarity tends to zero.

In many cases, numeric values in the input texts may include special characters (e.g., prices and phone numbers). Thus, prior to the application of the NM function, these characters are removed and the remaining numbers are concatenated. This process is called *Normalization*. For instance, the string "412-638-7273" is normalized to form a numeric value 4126387273 that can be applied to the function NM. Normalization is also performed over numeric values that occur in the knowledge base.

3.3.3 Attribute Value Format

In this approach, the common style often used to represent values of some attributes also considered as a feature. Content-based feature functions based on this aspect evaluate how likely are sequences of symbols forming a string in the input text. For this, typical sequences of symbols occurring on the values of an attribute in the knowledge base are learned. By using such features, it is possible to capture specific formatting properties of URLs, e-mails, telephone numbers, etc. In early methods, these features were learned over training data (Agichtein and Ganti 2004; Mansuri and Sarawagi 2006). In the presented approach, it is shown that is possible to compute them over data available in the knowledge base.

Again, let v_A be the set of values available for an attribute A in the knowledge base. The presented approach automatically learns a sequence Markov model m_A that captures the format style of the values in v_A. This model is similar to the inner HMM used in (Borkar et al. 2001) and is also applied to capture the format of values as a state feature.

For that, each value of v_A is first tokenized on white-spaces. Using a taxonomy proposed in (Borkar et al. 2001), the value is encoded as a sequence of *symbol masks* or simply *masks*. A mask is a character class identifier, possibly followed by a quantifier. Figure 3.3 illustrates an example of a taxonomy of symbols. As it can be noticed, at the top most level there is no distinction among symbols, at the next level, they are divided into *Numbers* and *Words*. The masks used to encode the input textual values are on the leaves of the taxonomy.

Then, the model m_A is generated based on these masks, so that each node n corresponds to a mask that represents the values of v_A. An edge e between nodes n_i and n_j is built if n_i is followed by n_j in the masks. Thus, each value in v_A can be described by a path in m_A.

To illustrate this concept, consider the knowledge base presented in Fig. 3.2. As stated earlier, it is possible to build a Markov Model for each attribute, **Neighborhood**, **Street**, **Bathrooms**, and **Phone**. Encoding the values of the attribute **Street** according to a predefined taxonomy of symbol masks would give us the follow-

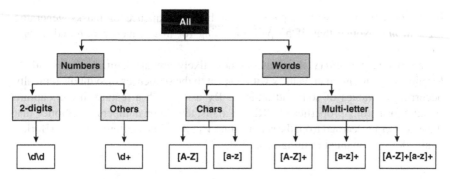

Fig. 3.3 Example of a Taxonomy of Symbols

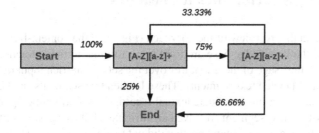

Fig. 3.4 A Markov model that represents the format of the values of the attribute Street

ing sequence of masks:"[A-Z][a-z]+ [A-Z][a-z]+." for representing the value *Regent St.*, "[A-Z][a-z]+ [A-Z][a-z]+." for *Morewood Ave.*, and "[A-Z][a-z]+ [A-Z][a-z]+. [A-Z][a-z]+" for *Square Ave. Park*. With this set of sequence of masks in hand, it is built the markov model that is depicted in Fig. 3.4.

To express the likelihood of sequences of masks in the model, it is defined the *weight* of an edge $\langle n_x, n_y \rangle$ as:

$$w(n_x, n_y) = \frac{\# \text{ of pairs } \langle n_x, n_y \rangle \text{ in } m_A}{\# \text{ of pairs } \langle n_x, n_z \rangle, \forall n_z \in m_A} \tag{3.4}$$

The Markov model depicted in the Fig. 3.4 shows that the values of the attribute *Street* always start with a word that has its first letter in uppercase and the following ones in lowercase. In 75 % of the values, this first word is followed by another word that finishes with a dot.

Now, let *s* be a candidate value. *s* can be encoded using the same symbol taxonomy as above. This results in a sequence of masks. This feature evaluates how similar a candidate value *s* is to the values in v_A with respect to their formats by computing

$$format(s, A) = \frac{\displaystyle\sum_{\langle n_x, n_y \rangle \in path(s)} w(n_x, n_y)}{|path(s)|} \tag{3.5}$$

where $path(s)$ represents a path formed by the sequence of masks generated for s in m_A. Notice that, if no path matching for this sequence is found in m_a, $format(s, A) = 0$.

Intuitively, $format(s, A)$ evaluates how likely are the sequences of symbols forming a given candidate value s with respect to the sequences of symbols typically occurring as values of some attribute A. By using such a feature, it can capture specific formatting properties of URLs, e-mails, telephone numbers, etc. Notice that the model m_A is learned from the set of values v_A only. Thus, differently from (Borkar et al. 2001), no manual training is needed.

3.4 Inducing Structure-Related Features

As described in the overview of the approach (Fig. 3.1), the content-based features. learned from the knowledge base are used to perform an initial extraction process. Consequently, the usage of these features over the set of data-rich input text snippets produce a set of labeled text segments. These labeled text segments can be arranged into groups that constitute candidate textual records. It is worth noticing that, at this point, most of the text segments received an attribute label using only content-based features, but there are some segments that did not receive any label, which are called *unmatched*.

Consider a candidate record $R = s_1, \ldots, s_r$, where each $s_i (1 \leq i \leq r)$ is a candidate value. Also, consider an attribute A and let ℓ_A be a label used for this attribute. Then, for any candidate value s_i, it can be computed the value of a feature function $f^k(s_i, A, R)$. Function f^k returns a real number that measures the likelihood of a segment labeled ℓ_A to occur in the same place as s_i in R. Thus, the value of f^k is related to the structure of R.

Differently from the content-based features used so far, which are only domain-dependent, structure-based features such as f^k depend on the particular organization of the candidate values within the input text. This means that these features are *source-dependent*.

State-of-the-art information extraction methods (Cortez et al. 2010, 2011; Mansuri and Sarawagi 2006; Zhao et al. 2008) usually use two types of structure-based feature. The first type considers the absolute position of the text segment or token to be evaluated and the second one considers its relative position, i.e., its occurrence between other segments or tokens in the input text. For computing such features, it is common to build a graph model that represents the likelihood of transitions within the input text (or other input texts from the same source).

In most CRF-based methods, this model is built from training data, which consists of a set of delimited records manually labeled taken from the same input (Mansuri and Sarawagi 2006). In (Cortez et al. 2010, 2011; Zhao et al. 2008), the model is built in an unsupervised way during the extraction process itself. While in (Zhao et al. 2008) a fixed order, learned from a sample, is assumed for the attributes in the input text, in the presented approach the model is built using all records available in the input text and no fixed order is assumed. More specifically, it is built a probabilistic HMM-like graph model called PSM (*Positioning and Sequencing Model*).

In this case, a PSM consists of: (1) a set of states $L = \{begin, \ell_1, \ell_2, \ldots, \ell_n, end\}$ where each state ℓ_i corresponds to a label assigned to a candidate value in the structure-free labeling step, (2) a matrix T that stores the probability of observing a transition from state ℓ_i to state ℓ_j, and (3) a matrix P that stores the probability of observing a label ℓ_i in the set of candidate labels that occupies the k-th position in a candidate record.

Matrix T, which stores the transition probabilities, is built using the ratio of the number of transitions made from state ℓ_i to state ℓ_j in a candidate record to the total number of transitions made from state ℓ_i in all known candidate records. Thus, each element $t_{i,j}$ in T is defined as:

$$t_{i,j} = \frac{\text{\# of transitions from } \ell_i \text{ to } \ell_j}{\text{Total \# of transitions out of } \ell_i} \tag{3.6}$$

Matrix P, which stores the position probabilities, is built using the ratio of the number of times a label ℓ_i is observed in position k in a candidate record to the total number of labels observed in candidate values that occupy position k in all known candidate records. Thus, each element $p_{i,k}$ in P is defined as:

$$p_{i,k} = \frac{\text{\# of observations of } \ell_i \text{ in } k}{\text{Total \# of candidate values in } k} \tag{3.7}$$

By using Eqs. 3.6 and 3.7, matrices T and P are built to maximize the probabilities of the sequencing and the positioning observed for the attribute values, according to the labeled text segments in the output of labeled using only the content-based features. This follows the Maximum Likelihood approach, commonly used for training graphical models (Borkar et al. 2001; Sarawagi 2008).

In practice, building matrices T and P involve performing a single pass over the output of the usage of the content-based features. Notice that text segments left unmatched are discarded when building these matrices. Obviously, possible mismatched text segments will be used to built the PSM, generating spurious transitions. However, as the number of mismatches is rather small, as demonstrated in the experiments, they do not compromise the overall correctness of the model.

Figure 3.5 shows an example of the PSM built for a set of data-rich input text containing classified ads. As it can be seen, the graph represents not only information on the sequencing of labels assigned to candidate values, but also on the positioning of candidate values in the input text. For instance, in this example, input texts are more likely to begin with text segments labeled as **Neighborhood** than with segments labeled as **Street**. Also, there is a high probability that text segments labeled as **Phone** occurring after segments labeled as **Bedrooms**.

Let s_k be a candidate value in a candidate record $R = \ldots, s_k, \ldots$ for which a label ℓ_i corresponding to an attribute A_i is to be assigned. Also, suppose that in R the candidate value next to s_k is labeled with ℓ_j corresponding to an attribute A_j.

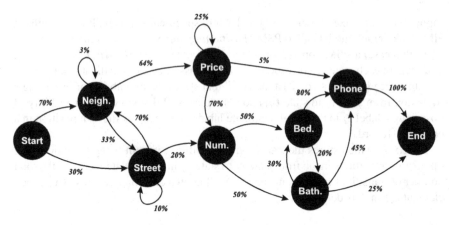

Fig. 3.5 Example of a PSM

Then, using Eqs. 3.6 an 3.7, it can be computed that the two structure-based features considered, i.e, the *sequencing feature* and the *positioning feature*, respectively as:

$$seq(s_k, A_i, R) = t_{i,j} \text{ and } pos(s_k, A_i, R) = p_{i,k} \qquad (3.8)$$

3.5 Automatically Combining Features

Given a candidate value s, the decision on which attribute label must be assigned to it takes into account different features. To combine these features, it is assumed that they represent the probability of the candidate value s to occur as a value of the attribute A domain, according to the knowledge base. If it is assumed that these features exploit different properties of the attribute A domain, it can be said they are independent, what allows us to combine them by means of the Bayesian disjunctive operator *or*, also known as *Noisy-OR-Gate* (Pearl and Shafer 1988).

It has considered several alternatives for such combination, including the use of machine learning approaches, such as SVM (Joachims 1998) and Genetic Programming (Fan et al. 2004), linear combination of values, and the use of a Bayesian Network framework. The use of machine learning is certainly an attractive alternative, but has the disadvantage of requiring a training, which would hamper the use of the presented approach in the application scenarios considered. The linear combination approach has provided fairly good results, but the quality of the assignments was a bit worse than the one obtained by the Bayesian framework.

Another alternative approach for the combination of features would be using some explicit optimization processes as those used in methods based on HMM and CRF. In HMM-based methods (Borkar et al. 2001), a Viterbi algorithm is used for finding the most likely path in a given HMM. Similarly, CRF-based methods (Mansuri and

Sarawagi 2006, 2008; Zhao et al. 2008) find the weight of each feature using iterative scaling algorithms (Lafferty et al. 2001). These optimization processes are very time consuming and the results obtained with them are similar to those achieved using the Bayesian approach adopted.

Although not using learning or optimization approaches can, in theory, lead to suboptimal results; experiments demonstrate that this combination approach works very well in practice. In addition, it has the advantage of speeding up the extraction process, as it is shown in the experiments presented in Sects. 4.5.2 and 5.8.6. Indeed, the hypothesis over the independence of the features gives to the presented a high level of automation and flexibility.

In the following it is described in detail how to combine content-based features and then, how to combine content-based and structure-based features.

3.5.1 Combining Content-Based Features

To combine content-based features, g^k, which are evaluated by feature functions of the form $g^k(s, A)$, as stated earlier, it is used a Bayesian disjunctive operator *or*, also known as *Noisy-OR-Gate* (Pearl and Shafer 1988), which is defined as:

$$or(p_1, \ldots, p_n) = 1 - ((1 - p_1) \times \ldots \times (1 - p_n)) \qquad (3.9)$$

where each p_i is a probability.

Thus, the final equation is:

$$\ell(s, A) = 1 - \left(\left(1 - g^1(s, A)\right) \times \ldots \times \left(1 - g^n(s, A)\right) \right) \qquad (3.10)$$

Informally, by using the disjunctive operator it is assumed that any of the features is likely to determine the labeling (i.e., significantly increase its final probability), regardless of other factors (Pearl and Shafer 1988). Function $\ell(s, A)$ is computed for each candidate value s in the input text for all attributes A of the same data type (i.e., text or numeric). Thus, s is labeled with a label representing the attribute that yielded the highest score according to this function.

3.5.2 Combining Structure-Based and Content-Based Features

Once the presented approach has induced structure-based features (Sect. 3.4), it can also combine them with content features in order to achieve the final extraction result. In this case, given a candidate value s, the decision on which label to assign to it can now consider the structure-based features f^j in addition to the content-based features g^k. As these features are also independent from the content-based ones, since they depend on the source, we again resort to the Bayesian Noisy-OR-Gate to combine all features as follows:

$$l\ell(s, R, A) = 1 - \left(\left(1 - g^1(s, A)\right) \times \ldots \times \left(1 - g^n(s, A)\right) \times \right.$$
$$\left. \left(1 - f^1(s, A, R)\right) \times \ldots \times \left(1 - f^m(s, A, R)\right) \right) \qquad (3.11)$$

Function $\ell(s, R, A)$ is computed, for each candidate segment s of all candidate records R in the input text, for all attributes A of the same data type (i.e., text or numeric). Thus, s is labeled with a label representing the attribute that yielded the highest score according to ℓ.

3.6 Unsupervised Extraction Methods

In the following chapters, it presented the unsupervised information extraction by text segmentation methods that were developed based on the presented unsupervised approach.

Chapter 4 describes an extraction method called *ONDUX* (On Demand Unsupervised Information Extraction) (Cortez and da Silva 2010; Cortez et al. 2010; Porto et al. 2011). *ONDUX* relies on content-based features, learned from knowledge bases and structured-based features, directly learned *on-demand* from test data, to perform extraction over unstructured textual records.

Chapter 5 describes an other information extraction method that was developed called *JUDIE* (Joint Unsupervised Structure Discovery and Information Extraction) (Cortez et al. 2011). Similarly, to *ONDUX*, *JUDIE* also relies on content-based and structured-based features to perform the extraction task. But, unlike other existing extraction methods, *JUDIE* is capable of detecting the structure of each individual record being extracted without any user assistance. This feature unique of *JUDIE* is accomplished by a novel Structure Discovery algorithm that was developed to tackle this problem.

The presented approach was also exploited by *iForm* (Toda et al. 2009, 2010), a method that is able to deal with the Web form filling problem. Chapter 6 presents an overview of this method and show how it exploits values that were previously submitted to Web forms to learn content-based features, which are then used to extract values from unstructured text. As in the form filling setting the usage of structure-based features is not possible, *iForm* relies only on content-based features.

All of these methods were developed, implemented, and evaluated considering different experimental datasets. All the experiments performed to validate them are described in the following chapters.

References

Agichtein, E., & Ganti, V. (2004). Mining reference tables for automatic text segmentation. *Proceedings of the ACM SIGKDD International Conference on Knowledge Discovery and Data Mining* (pp. 20–29). Seattle, USA.

Agrawal, S., & Chaudhuri, S. (2003). Automated ranking of database query results. In *Proceedings of the CIDR Biennial Conference on Innovative Data Systems Research*, Asilomar, USA.

Borkar, V., Deshmukh, K., & Sarawagi, S. (2001). Automatic segmentation of text into structured records. *Proceedings of the ACM SIGMOD International Conference on Management of Data Conference* (pp. 175–186). Santa Barbara, USA.

Chiang, F., Andritsos, P., Zhu, E., & Miller, R. (2012). Autodict: Automated dictionary discovery. *Proceedings of the IEEE ICDE International Conference on Data Engineering* (pp. 1277–1280). Washington, USA.

Cortez, E., da Silva, A., Gonçalves, M., & de Moura, E. (2010). ONDUX: On-demand unsupervised learning for information extraction. *Proceedings of the ACM SIGMOD International Conference on Management of Data Conference* (pp. 807–818). Indianapolis, USA.

Cortez, E., da Silva, A., Gonçalves, M., Mesquita, F., & de Moura, E. (2007). FLUX-CIM: flexible unsupervised extraction of citation metadata. *Proceedings of the ACM/IEEE JCDL Joint Conference on Digital Libraries* (pp. 215–224). Vancouver, Canada.

Cortez, E., & da Silva, A. S. (2010). Unsupervised strategies for information extraction by text segmentation. *Proceedings of the SIGMOD PhD Workshop on Innovative Database Research* (pp. 49–54). Indianapolis, USA.

Cortez, E., da Silva, A. S., de Moura, E. S., & Laender, A. H. F. (2011). Joint unsupervised structure discovery and information extraction. *Proceedings of the ACM SIGMOD International Conference on Management of Data Conference* (pp. 541–552). Athens, Greece.

Fan, W., Gordon, M., & Pathak, P. (2004). Discovery of context-specific ranking functions for effective information retrieval using genetic programming. *IEEE Transactions on knowledge and Data Engineering, 16*(4), 523.

Joachims, T. (1998). Text categorization with support vector machines: Learning with many relevant features. *Proceedings of the European Conference on Machine Learning* (pp. 137–142). Chemnitz, Germany.

Lafferty, J., McCallum, A., & Pereira, F. (2001). Conditional random fields: Probabilistic models for segmenting and labeling sequence data. *Proceedings of the ICML International Conference on Machine Learning* (pp. 282–289). Williamstown, USA.

Mansuri, I. R., & Sarawagi, S. (2006). Integrating unstructured data into relational databases. *Proceedings of the IEEE ICDE International Conference on Data Engineering* (pp. 29–41). Atlanta, USA.

Mesquita, F., da Silva, A., de Moura, E., Calado, P., & Laender, A. (2007). LABRADOR: Efficiently publishing relational databases on the web by using keyword-based query interfaces. *Information Processing and Management, 43*(4), 983–1004.

Pearl, J., & Shafer, G. (1988). *Probabilistic reasoning in intelligent systems: networks of plausible inference*. San Francisco: Morgan Kaufmann Publishers Inc.

Porto, A., Cortez, E., da Silva, A. S., & de Moura, E. S. (2011). *Unsupervised information extraction with the ondux tool*. Florianpolis: In Simpsio Brasileiro de Banco de Dados.

Salton, G., Wong, A., & Yang, C. (1975). A vector space model for automatic indexing. *Communications of the ACM, 18*(11), 613–620.

Sarawagi, S. (2008). Information extraction. *Foundations and Trends in Databases, 1*(3), 261–377.

Serra, E., Cortez, E., da Silva, A., & de Moura, E. (2011). On using wikipedia to build knowledge bases for information extraction by text segmentation. *Journal of Information and Data Management, 2*(3), 259.

Toda, G., Cortez, E., da Silva, A. S., & de Moura, E. S. (2010). A probabilistic approach for automatically filling form-based web interfaces. *Proceedings of the VLDB Endowment, 4*(3), 151–160.

Toda, G., Cortez, E., Mesquita, F., da Silva, A., Moura, E., & Neubert, M. (2009). Automatically filling form-based web interfaces with free text inputs. *Proceedings of the WWW International World Wide Web Conferences* (pp. 1163–1164). Madrid, Spain.

Toda, G. A., & da Silva, A. S. (2006). *Um Mtodo Probabilstico para o Preenchimento Automtico de Formulrios Web a Partir de Textos Ricos em Dados*. Universidade Federal do Amazonas.

Zhao, C., Mahmud, J., & Ramakrishnan, I. (2008). Exploiting structured reference data for unsupervised text segmentation with conditional random fields. *Proceedings of the SIAM International Conference on Data Mining* (pp. 420–431). Atlanta, USA.

Chapter 4
ONDUX

Abstract This chapter presents *ONDUX* (On Demand Unsupervised Information Extraction) a method that relies on the presented unsupervised approach to deal with the Information Extraction by Text Segmentation problem. *ONDUX* was first presented in Cortez et al. (2010) and in Cortez and da Silva (2010). Following, a tool based on *ONDUX* was presented in Porto et al. (2011). As other unsupervised IETS approaches, *ONDUX* relies on information available on pre-existing data, but, unlike previously proposed methods, it also relies on a very effective set of content-based features to bootstrap the learning of structure-based features. More specifically, structure-based features are exploited to disambiguate the extraction of certain attributes through a reinforcement step. The reinforcement step relies on sequencing and positioning of attribute values directly learned *on-demand* from test data. In the following, it is presented an overview of *ONDUX* and describe the main steps involved in its functioning. Next, each step is discussed in turn with detail. It also reported an experimental evaluation of *ONDUX* presenting its performance in different datasets and domains. Finally, it is described as a tool that implements the *ONDUX* method.

Keywords Information extraction · Unsupervised approach · Text segmentation · Databases · Knowledge bases · On demand

4.1 Overview

Consider an input string I representing a real classified ad such as the one presented in Fig. 4.1a. As stated in Chap. 1, the IETS problem consists in segmenting I in such a way that each segment s receives a label ℓ corresponding to an attribute a_ℓ, where

---

This chapter has previously been published as Cortez et al. (2010); reprinted with permission.

E. Cortez and A. S. da Silva, *Unsupervised Information Extraction by Text Segmentation*,   33
SpringerBriefs in Computer Science, DOI: 10.1007/978-3-319-02597-1_4,
© The Author(s) 2013

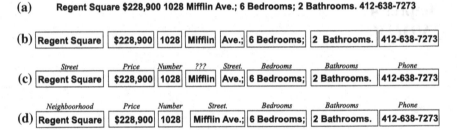

Fig. 4.1 Example of an extraction process on a classified ad using *ONDUX*

s represents a value in the domain of a_ℓ. This is illustrated in Fig. 4.1d, which is an example of the outcome produced by *ONDUX*.

Similar to previous approaches (Agichtein and Ganti 2004; Zhao et al. 2008), in *ONDUX* it is to used attribute values that come from pre-existing data sources from each domain (e.g. addresses, bibliographic data, etc.) to label segments in the input text. These values are used to form domain-specific Knowledge Bases, according to the definition in Sect. 3.2.

The *ONDUX* first step is called *Blocking*. In this step, the input string is roughly segmented into units called *blocks*. Blocks are simply sequences of terms (words) that are likely to form a value of an attribute. Thus, although terms in a block must all belong to a same value, a single attribute value may have terms split among two or more blocks. This concept is illustrated in Fig. 4.1b. Observe that the blocks containing the terms "Mifflin" and "Ave" are parts of the same value of attribute Street.

Next, in the *Matching* step, blocks are associated with attribute labels using the content-based features (Sect. 3.3) that were learned from a knowledge base. By the end of the Matching step, each block is *prelabeled* with the name of the attribute for which the best match was found.

We notice that the Blocking and Matching steps alone are enough to correctly label the large majority of the segments in the input string. Indeed, experiments with different domains, which were performed and reported here, have shown that blocks are correctly prelabeled in more than 70 % of the cases. This is illustrated in Fig. 4.1c in which the Matching was able to successfully label all blocks except for the one containing the terms "Regent Square" and "Mifflin."

Problems such as these are likely to occur in two cases. The first, *Mismatching*, happens when two distinct attributes have domains with a large intersection. For instance, when extracting from scientific paper headings, values from attributes Title and Keywords have usually several terms (words) in common. In the running example, as shown in Fig. 4.1c, "Regent Square" was mistakenly labeled with Street instead of Neighborhood. The second, *Unmatching*, happens when the content-based features used are not able to determine any label to a given block, as the case of the block containing the term "Mifflin" in Fig. 4.1c.

To deal with such problems, the *ONDUX* method includes a third step called *Reinforcement* in which the outcome of the Matching step is explored to automatically induce structure-based features (Sect. 3.4) and, the Matching step is reinforced by taking into consideration the positioning and the sequencing of labeled blocks in the input texts. In the following, it presented the details of each step described above.

4.2 Blocking Step

The first step of *ONDUX* consists of splitting an input string into substrings called *blocks*. In the presented method, blocks are considered as sequences of terms that will compose a single value of a certain attribute. In Fig. 4.1b the blocks identified in the input string example are inside rectangles.

The blocking process is based on the co-occurrence of terms in a same attribute value according to the knowledge base. This process is described in Algorithm 1.

Let I be an input string. Initially, terms are extracted from I based on the occurrence of white spaces in the string, being special symbols and punctuation simply discarded (Line 1).

Next (Lines 7–15), blocks are built as follows: if the current term (say, t_{j-1}) and next term (say, t_j) are known to co-occur in some occurrence in the knowledge base, then t_j will compose the same block as t_{j-1}. Otherwise, a new block will be built for t_j. This process is repeated until all terms of I are assigned to a block. Notice that terms that do not occur in the knowledge base always form a block alone.

Algorithm 1 Blocking

1: I : Input Text
2: $K = \{\langle a_1, O_1 \rangle, \dots, \langle a_n, O_n \rangle\}$: knowledge base
3: $T : \langle t_0, \dots, t_n \rangle \leftarrow ExtractTerms(I)$
4: $B_0 \leftarrow \dots \leftarrow B_n \leftarrow \emptyset$ {Initialize blocks}
5: $B_0 \leftarrow B_0 \cup \langle t_0 \rangle$; {Builds the first block}
6: $i = 0, \ j = 1$;
7: **repeat**
8:     $C \leftarrow \{\langle a_k, O_k \rangle \in K, \ o_x \in O_k \mid t_{j-1}, t_j \in o_x\}$
9:     **if** $C = \emptyset$ **then**
10:         $\{t_{j-1}$ and t_j do not co-occur$\}$
11:         $i \leftarrow i + 1$; {Next block}
12:     **end if**
13:     $B_i \leftarrow B_i \cup \langle t_j \rangle$; {Adds t_j to the current block}
14:     $j + +$; {Next term}
15: **until** $j = n$

According to the knowledge base presented in Fig. 3.2 (Sect. 3.2), terms "Regent" and "Square" co-occur as values of the attribute Neighborhood. Thus, as shown in Fig. 4.1b, these terms are in the same block, i.e., the first block in the figure.

4.3 Matching Step

The Matching step consists in associating each block generated in the Blocking step with an attribute represented in the knowledge base. For this, it is used in the content-based features described in Sect. 3.3. These features are used to determinate the attribute that the block is more likely to belong to. The specific content-based feature that will be used to match a block is chosen by a simple test over the terms composing this block to define a data type. It is considered four distinct types of data with a corresponding content-based feature: *text*, *numeric*, *urls*, and *email*.

For the matching of textual values, *ONDUX* relies on the Attribute Vocabulary feature described in Sect. 3.3.1. This feature exploits the common vocabulary often shared by values of textual attributes (e.g., neighborhood and street names, author names, recipe ingredients, etc.). For the matching of numeric values *ONDUX* relies on the Attribute Value Range feature described in Sect. 3.3.2. The Attribute Value Range feature specifically deals with numeric attributes using the average and the standard deviation of the values of numeric attributes available on the knowledge base. For matching URLs and e-mails values *ONDUX* applies simple binary functions using regular expressions, which identify each specific format and return true or false.

Despite its simplicity, the use of content-based features adopted to label blocks is by itself a very effective way of labeling segments in the input text. Indeed, experiments with different domains, which were performed and reported here, show that blocks are correctly prelabeled in more than 70 % of the cases.

In Fig. 4.1c it is shown that the result obtained after the matching step for the running example. As can be noticed, almost all blocks were assigned to a proper attribute, except for the following cases: (1) the block containing "Mifflin" was left unmatched and (2) the block containing "Regent Square" was mistakenly assigned to Street, instead of being assigned to Neighborhood. To deal with both cases, the *ONDUX* method includes a third step, *Reinforcement*, which is discussed in the following section.

4.4 Reinforcement Step

The Reinforcement step consists in revising the prelabeling made by the Matching step over the blocks. More specifically, unmatched blocks are labeled and mismatched blocks are expected to be correctly relabeled. It can be noticed that in this context, the term Reinforcement is used in a sense slightly different from the traditional Reinforcement Learning technique (Kaelbling et al. 1996). Indeed, in this case this step not only reinforces the labeling performed by the Matching step, but also revises and possibly corrects it.

As the prelabeling of blocks performed in the Matching step has a high accuracy (as demonstrated by the experiments), this prelabeling can be used to automatically induce structure-based features (Sect. 3.4), which are related to the sequencing and positioning of attribute values in input texts. Notice again that these features are learned *on-demand* from each set of input text with no need for human training nor assumptions regarding a particular order of attribute values.

For computing such structure-based features, it is common to use a graph model that represents the likelihood of attribute transitions within the input text (or any other input text from the same source). It is used as a probabilistic HMM-like graph model that is called PSM (Positioning and Sequencing Model). The process of automatically inducing structure-based features and building the PSM model is explained in detail in Sect. 3.4. After generating the PSM, the estimated probabilities are used to perform label reinforcement, as discussed in the following section.

On the Matching step, the labeling of a block was based entirely on the content-based features as described in Sect. 4.3. However, after building the PSM, the decision on what label to assign to a block can also take into account the structure-based features of the text inputs.

To combine the content-based features and the structure-based features, *ON-DUX* relies on a combination strategy described in Sect. 3.5. Notice that there will be no unmatched blocks after this process. Once all blocks are labeled, contiguous blocks with a same label are merged. Thus, each block would correspond to a single attribute value. This is illustrated in the running example in Fig. 4.1d, in which all blocks are correctly assigned to the attributes. The first block, which was wrongly labeled in the Matching phase, has been now correctly assigned to the Neighborhood attribute. Also, the unmatched block containing the term "Miffin" now composes a value of attribute Street.

4.5 Experimental Evaluation

In this section, it is presented as an evaluation of *ONDUX* using a variety of real datasets to show that this is a robust, accurate, and efficient unsupervised method for IETS. It has first described the experimental setup, including experimental data and the metrics used. Then, it is reported results on extraction quality and performance over all distinct datasets.

4.5.1 Setup

In the experiments, we compare *ONDUX* with an unsupervised version of CRF. This version was developed by adapting the publicly available implementation of CRF by Sunita Sarawagi,[1] according to what is described in Zhao et al. (2008). This version is called *U-CRF*. We believe that *U-CRF* represents the most suitable baseline for comparing with *ONDUX*, as it delivers top performance while at the same time it

[1] http://crf.sourceforge.net/

does not require user-provided training. Although the Extended Semi-markov CRF presented in Mansuri and Sarawagi (2006) could have been used as baseline, since it relies mostly on features learned from a pre-existing dataset, it also uses a small portion of manually trained data. Moreover, Zhao et al. (2008) improves on Mansuri and Sarawagi (2006) results. However, since the first baseline assumes, it shall be seen in more detail later, that the order of the text sequences to be extracted is fixed, it was also included that the standard CRF model (Lafferty et al. 2001) (called *S-CRF*), does not have this limitation at all but requires manually labeled training data. Obviously, *S-CRF* is only used as a baseline for cases in which we have the training data. Using the two baselines, also allows us to compare the strengths of each of these models against the presented unsupervised approach.

As for the configuration of *U-CRF* and *S-CRF*, it was deployed that the same features are described in Zhao et al. (2008) and in Lafferty et al. (2001). Overall, these are standard features available on the publicly CRF implementation, e.g., dictionary features, word score functions, transition features, etc., plus, in the case of *U-CRF* the set of heuristic rules for using negative examples proposed in Zhao et al. (2008). Although the basic CRF model is flexible enough to allow features to be tailored for specific extractions tasks, in all experiments we have used the same configuration for *U-CRF* and *S-CRF*. This is to ensure a fair comparison since it is assumed that no specific adjustments were necessary for *ONDUX* to be used in the experiments.

As required by *U-CRF*, a batch of the input strings is used to infer the order of the attribute values. Based on the information provided in Zhao et al. (2008), this batch is composed of 10 % of the input strings in all cases.

4.5.1.1 Experimental Data

The data sources used to generate the knowledge bases for *ONDUX*, the references tables for *U-CRF* and the training data for *S-CRF* as well as the test datasets used in the experiments are summarized in Table 4.1.

It was tried to use the same datasets and data sources explored by the baselines, when these were publicly available. In the case of restricted data sources or datasets, we tried to obtain similar public versions on the same domains.

Indeed, in most cases the data sources and the test datasets that were used came from public available sources used for the empirical analysis of information extraction methods. This is the case of *Bigbook* and *Restaurants*, from the RISE

Table 4.1 Domains, data sources, and test datasets are used in the experiments

Domain	Source	Attribute	Record	Dataset	Attribute	Inputs
Addresses	*BigBook*	5	2000	*BigBook*	5	500–2000
				Restaurants	4	250
Bibliographic Data	*CORA*	13	350	*CORA*	13	150
	PersonalBib	7	395			
Classified Ads	*Folha On-line*	5–18	125	*Web Ads*	5–18	500

repository (Muslea 2012), the *CORA* collection (McCallum 2012), and the *PersonalBib* dataset (Mansuri and Sarawagi 2006). It is important to notice that in the case of *BigBook* and *CORA*, the knowledge bases and the reference tables were built from sets of records already extracted by third-parties and those are completely disjoint (i.e., have no common entries) from the strings in the test datasets were used in the experiments.

Data on the *Classified Ads* domain were obtained directly from the Web. For building the knowledge base, it was collected that data from an on-line database available from *Folha On-line*, a popular Brazilian newspaper site. The test dataset *Web Ads* is formed by unstructured strings containing ads from other five Brazilian newspaper sites. Each website bares a distinct classified ads format, e.g., in terms of attribute values order and positioning. Moreover, the number of distinct attribute occurrences in each instance varies from 5 to 18. These properties result in a high level of heterogeneity in the test instances.

4.5.1.2 Metrics for Evaluation

In these experiments it was evaluated that the extraction results obtained after the Matching and Reinforcement steps were discussed in Sect. 4.1. We aimed at verifying how each step contributes to the overall effectiveness of *ONDUX*. In the evaluation we used the well-known precision, recall, and F-measure metrics, but all tables report only F-measure values.

Let B_i be a reference set and S_i be a test set to be compared with B_i. We define precision (P_i), recall (R_i), and F-measure (F_i) as:

$$P_i = \frac{|B_i \cap S_i|}{|S_i|} \qquad R_i = \frac{|B_i \cap S_i|}{|B_i|} \qquad F_i = \frac{2(R_i.P_i)}{(R_i + P_i)} \tag{4.1}$$

For all reported comparisons with U-CRF, it used the Student's T-test (Anderson and Finn 1996) for determining if the difference in performance was statistically significant. In all cases, we only drew conclusions from results that were significant in, at least, 5 % level for both tests. Nonsignificant values are omitted.

Also, we run each experiment five times, each time selecting different samples for building the knowledge base and for testing. For all the experiments we performed, we report the average of the results obtained in each of the five runs.

4.5.2 Extraction Evaluation

4.5.2.1 Blocking Results

The first reported result aims at verifying in practice the strategy formulated for the Blocking step, that is, the blocking strategy only generates blocks in which all terms

Table 4.2 Results of experiments on the blocking step

Dataset	Source	% Same	% Unknown
BigBook	BigBook	94.13	5.34
Restaurants	BigBook	92.17	7.42
CORA	CORA	80.91	18.88
CORA	PersonalBib	78.00	19.47
Web Ads	Folha On-line	87.13	12.32

belong to a unique attribute. Thus, it is measured how homogeneous each generated block is.

Table 4.2, column "% Same" shows that in all test datasets a large percentage of blocks contain terms found in the values of the same attribute according to the knowledge base. Column "% Unknown" shows the percentage of blocks with terms not represented in the knowledge base. As pointed out in Sect. 4.2, such blocks always contain a single term. It is noticed that in all cases the percentage of heterogeneous blocks, that is, those that are not homogeneous nor unknown is rather small, which is less than 3 %. Thus, it can be concluded that the blocking strategy behaves as expected.

It is worth mentioning that the high percentage of unknown blocks in the CORA dataset is caused by the diversity of terms that is normally found in the scientific paper metadata, specially in the Title attribute. As it shall be seen latter, despite this, ONDUX shows an excellent performance on this dataset.

4.5.2.2 Attribute-Level Results

To demonstrate the effectiveness of the method in the whole extraction process, it is evaluated that its extraction quality by analyzing, for each attribute, if the (complete) values assigned by the ONDUX method to this attribute are correct. In what follows it is shown the results for the three domains considered: Addresses, Bibliographic Data, and Classified Ads.

Addresses Data Domain

Table 4.3 shows the results for the attribute level extraction over the BigBook dataset using the BigBook data source. Recall that, although the same collection has been used, the dataset and the data source are disjoint. This is the same experiment reported in Zhao et al. (2008), and it included here for completeness and to validate the baseline implementation. The BigBook dataset follows the assumption made by Zhao et al. (2008), according to which "a batch of text sequences to be segmented shares the same total attribute order." This is called single total attribute order assumption.

Table 4.3 Extraction over the *BigBook* dataset using data from the *BigBook* source

Attribute	S-CRF	U-CRF	ONDUX	
			Matching	Reinforcement
Name	0.997	0.995	0.928	0.996
Street	0.995	0.993	0.893	0.995
City	0.986	0.990	0.924	**0.995**
State	0.999	0.999	0.944	1.000
Phone	0.992	0.988	0.996	**1.000**
Average	0.994	0.993	0.937	**0.997**

In Table 4.3, values in boldface indicate a statistically superior result with at least 95 % confidence. Starting by the comparison between the unsupervised methods, it can be seen that the results of both *U-CRF* and *ONDUX* after the reinforcement are extremely high for all attributes (higher than 0.988 for all attributes). However, the results of the *ONDUX* method are statistically superior than those of *U-CRF* in at least two attributes (i.e., City and Phone and are statistically tied in the other three attributes. Another important aspect is the importance of the reinforcement step which produced gains of more than 5 % over very strong results. A closer look at this gain, reveals that it is mostly due to recall, which improved more that 9 %, while the precision improved only 2 %, on average. This is in accordance with our hypothesis regarding the high precision of the Matching step. The Reinforcement step plays the role of "filing the gaps," and therefore, improving recall. Notice that the U-CRF results are very similar to those reported in Zhao et al. (2008), thus further validating the baseline implementation.

Since in this case we have manually labeled data in the *BigBook* dataset, we were also able to compare the unsupervised methods with *S-CRF*. In this case, the results of both CRF-based methods are very close, and the conclusions are similar to the one described before. This also shows that the supervised method, in this particular dataset, could not take much advantage of the training data besides what U-CRF was able to learn from the reference tables.

This experiment was repeated using the *Restaurants* dataset as the test dataset. The motivation is to show that IETS approaches based on previously known data such as *ONDUX* and *U-CRF* are capable of learning and using source independent features from these data. In this case, as well as in the others in which the source is different from the test dataset, the comparison with the *S-CRF* does not make sense, since, for this method to work, the learning data has to present a similar distribution as the test data. The *Restaurants* dataset has the same attributes as the *BigBook* one, except for the State attribute. The single total attribute order assumption also applies here. The results are reported in Table 4.4.

Again, both *U-CRF* and *ONDUX* achieved good results for all attributes, higher than 0.942 for all attributes. *ONDUX* had a statistically significant advantage on attributes Name and Phone, while statistical ties were observed for attributes Street and City.

Table 4.4 Extraction over the *Restaurants* dataset using data from the *BigBook* source

| Attribute | U-CRF | ONDUX | |
		Matching	Reinforcement
Name	0.942	0.892	**0.975**
Street	0.967	0.911	0.982
City	0.984	0.956	0.987
Phone	0.972	0.982	**0.992**
Average	0.966	0.935	**0.984**

Table 4.5 Extraction over the *CORA* dataset using data from the *CORA* source

| Attribute | S-CRF | U-CRF | ONDUX | |
			Matching	Reinforcement
Author	0.936	0.906	0.911	**0.960**
Booktitle	0.915	0.768	0.900	0.922
Date	0.900	0.626	0.934	**0.935**
Editor	0.870	0.171	0.779	**0.899**
Institution	**0.933**	0.350	0.821	0.884
Journal	0.906	0.709	0.918	**0.939**
Location	0.887	0.333	0.902	0.915
Note	0.832	0.541	0.908	**0.921**
Pages	**0.985**	0.822	0.934	0.949
Publisher	0.785	0.398	0.892	**0.913**
Tech	0.832	0.166	0.753	0.827
Title	**0.962**	0.775	0.900	0.914
Volume	0.972	0.706	0.983	0.993
Average	0.901	0.559	0.887	**0.921**

Bibliographic Data Domain

The next set of experiments was performed using the *CORA* test dataset. This dataset includes bibliographic citations in a variety of styles, including citations for journal papers, conference papers, books, technical reports, etc. Thus, it does not follow the single total attribute order assumption made by Zhao et al. (2008). The availability of manually labeled data allowed us to include the *S-CRF* method in this comparison. A similar experiment is reported in Peng and McCallum (2006). Because of this, we have to generate the knowledge base and the reference tables for *U-CRF* using the same data available on the unstructured labeled records we used to train the standard CRF, also from the *CORA* collection. As always, this training data is disjoint from the test dataset. The results for this experiment are presented in Table 4.5.

First, notice that the good results obtained with the supervised CRF (*S-CRF*) are similar to those reported in the original experiment (Peng and McCallum 2006). In the case of *ONDUX*, although it is an unsupervised method, even superior results were achieved. Statistically superior results were obtained for 6 out of 13 attributes (results in boldface) and statistical ties were observed for other 4 attributes. The results with *U-CRF* were rather low, what is explained by the heterogeneity of the citations in the collections. While the manual training performed for *S-CRF* was able to capture this heterogeneity, *U-CRF* assumed a fixed attribute order. On the other hand, *ONDUX* was able to capture this heterogeneity through the PSM model, without any manual training.

Still on the Bibliographic data domain, we repeated the extraction task over the *CORA* test dataset, but this time, the previously known data came from the *PersonalBib* dataset. This dataset was used in a similar experiment reported in Mansuri and Sarawagi (2006). Again, our aim was demonstrated to the source independent nature of unsupervised IETS methods. Notice that not all attributes from *CORA* were present in *PersonalBib* entries. Thus, we only extracted attributes available on both of them. The results for this experiment are presented in Table 4.6. Notice that in this case we could not perform manual training, since the previously known data came directly from a structured source. Thus, we do not report results for S-CRF here.

The results for *ONDUX* and *U-CRF* are quite similar to those obtained in the previous experiments, with a large advantage for *ONDUX*, and for the reasons that were already discussed.

Classified Ads Domain

Finally, Table 4.7 presents the results for the experiments with the test dataset *Web Ads*. The knowledge base and the reference tables were built using structured data from the *Folha On-line* collection. In this table, the attribute *Others* corresponds to an amalgamation of a series of attributes present only in few ads such as Neighborhood, Backyard, Garden, etc. For this dataset, *ONDUX* outperforms *U-CRF* in about 5 % even before the Reinforcement step. After this step, the *ONDUX* method significantly outperforms the baseline in all attributes with an overall gain of more than 10 % in average. Recall that this is a very heterogeneous dataset bearing several distinct

Table 4.6 Extraction over the *CORA* dataset using data from the *PersonalBib* source

Attribute	U-CRF	ONDUX	
		Matching	Reinforcement
Author	0.876	0.733	**0.922**
Booktitle	0.560	0.850	**0.892**
Date	0.488	0.775	**0.895**
Journal	0.553	0.898	**0.908**
Pages	0.503	0.754	**0.849**
Title	0.694	0.682	**0.792**
Volume	0.430	0.914	**0.958**
Average	0.587	0.801	**0.888**

Table 4.7 Extraction over the *Web Ads* dataset using data from the *Folha On-line* source

Attribute	U-CRF	ONDUX	
		Matching	Reinforcement
Bedroom	0.791	0.738	**0.861**
Living	0.724	0.852	**0.905**
Phone	0.754	0.884	**0.926**
Price	0.786	0.907	**0.936**
Kitchen	0.788	0.776	**0.849**
Bathroom	0.810	0.760	0.792
Suite	**0.900**	0.853	0.881
Pantry	0.687	0.741	**0.796**
Garage	0.714	0.784	**0.816**
Pool	0.683	0.711	**0.780**
Others	0.719	0.777	**0.796**
Average	0.760	0.798	**0.849**

formats. The good results in this dataset highlights the robustness and the flexibility of the presented solution, even when compared to the closest competitor.

4.5.3 Dependency on Previously Known Data

An important question to address is to determine how dependent the quality of results provided by the unsupervised IETS methods studied is from the overlap between the previously known data and the text input. To study such dependency, it was performed that experiments to compare the behavior of *ONDUX* and *U-CRF* when varying the amount of terms given in the knowledge base or reference tables that overlap with the terms found in the input text. Recall that the entries in which these terms occur are used to form attribute occurrences in the knowledge base for *ONDUX*, and the reference tables for training *U-CRF*.

The experiments were performed using the *BigBook* dataset, which contains about 4000 entries. As mentioned earlier, this dataset came from the RISE repository (Muslea 2012). Thus, the knowledge base and the reference tables were build from sets of records already extracted, which are disjoint from the strings on the test datasets used from the same collections.

In the experiments, we vary the number of known terms that are shared between the previously known data and the input test sequence. We have also varied the number of input strings in the test sequence to check whether the amount of overlap necessary to obtain good results increase as the number of text inputs found in the test sequence also increases.

Figure 4.2 shows the results for four different sizes of test set, varying the number of text inputs present in the test set from 500 to 2000. The number of shared terms between the knowledge base and the test input sequence varies in all cases from 50 to 1000 terms, and the extraction quality is evaluated by means of F-measure.

An important information obtained from these four graphs is that the quality of results provided by the methods does not vary with the size of the test input for fixed amounts of shared terms. For instance, with an overlap of 250 terms, *ON-DUX* achieved 0.73 of F-measure for the test dataset of size 500 and 0.74 for the test dataset of size 1500. When taking an overlap of 100 terms, values are 0.66, 0.67, 0.68, and 0.64 for the test sizes 500, 1000, 1500, and 2000, respectively. These results indicate that, at least for this dataset domain, both *ONDUX* and *U-CRF* could keep good performance with a small amount of previously known data even for larger test sets. This behavior was expected, since both methods use the overlap to obtain statistics about the structure of the test input sequence. Once the number of term overlaps is large enough to allow the methods to compute such statistics, both methods are able to learn how to extract data from the test input sequence, no matter what is its size.

It can be also seen from the graphs that the total number of shared terms necessary to achieve good performance is also not prohibitive, since both methods were able to achieve high quality performance (more than 95 % in case of *ONDUX*) when taking only 750 terms of overlap for all the four sizes of test set studied. When looking to the smaller test sets, this overlap seems to be high when compared to the size

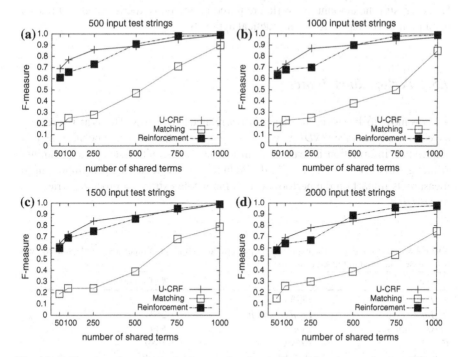

Fig. 4.2 F-Measure values obtained when varying the number of shared terms for four different sizes of datasets built from *BigBook*

of the test, but it does not need to increase as the test set increases. The number of records from the *BigBook* source required to obtain such overlap in the knowledge base was 162 in the results presented in Fig. 4.2d, about 8 % of the size of the test set (recall that these are disjoint sets). This overlap also represents about 14 % of vocabulary overlap between the knowledge base and the test set. These percentages are obviously higher for the smaller tests, since we still need 750 term overlaps to achieve about the same performance, but would tend to be zero for larger test sets.

A good question at this point is to know how practical to have hundred of terms in common between a reference set and a real data source for a system to extract information. To give a better idea about practical scenarios, let us consider all the combinations of data sources and datasets we tested in the experiments, where most collections were taken from previous experiments presented in literature.

The term overlap results found in the experiments with these combinations are depicted in Table 4.8. As it can be seem, except for the combination of PersonalBib as data source and CORA as dataset, in all the experiments performed the number of shared terms is higher than the amounts of shared terms found in Fig. 4.2, which allowed both *ONDUX* and *U-CRF* to achieve high level quality of results in the experiments. For instance, when using *BigBook* as data source and *Restaurants* as the test dataset, the number of shared terms is 2504. Of course, the overlap is not the unique factor to determine the performance of the methods and the amount of overlap required may vary according to other factors presented in the experiments. However, still the amount of overlap required by the two experimented methods is not a prohibitive aspect for their practical application.

4.5.4 Performance Issues

We now move to discuss performance issues related to *ONDUX*. This is an interesting aspect to analyze since *ONDUX* works *on-demand*, in the sense that positioning and sequencing information is learned from test instances, with not *a priori* training. Although this feature gives to *ONDUX* a high level of flexibility, it is important to measure its impact on the performance of the whole extraction process carried out by *ONDUX*.

Table 4.8 Term overlap in the experiments performed with all combinations of data sources and test datasets adopted in the experiments

Source	Dataset	# of shared terms
BigBook	*BigBook*	3667
BigBook	*LA Restaurants*	2504
PersonalBib	*CORA*	549
CORA	*CORA*	1089
Folha On-line	*Web Ads*	1184

Table 4.9 Time in seconds spent in each extraction task

Source	Dataset	U-CRF	ONDUX
BigBook	BigBook	316	23
BigBook	LA Restaurants	604	27
PersonalBib	CORA	317	21
CORA	CORA	194	17
Folha On-line	Web Ads	2746	19

Also in this aspect, we compare *ONDUX* with the baseline *U-CRF*. For this, we take into account training and test times. This is justified by the fact that every new extraction process carried out by *U-CRF* requires a new model to be learned from test instances.

The time figures reported here were collected for each one of the quality experiments presented earlier. For each specific task it was measured that the time in seconds was spent by each unsupervised extraction method. These results are presented in Table 4.9.

In spite of the on-demand process performed by *ONDUX*, the time spent on processing test instances is shorter than the time spent by *U-CRF*. In all experiments, it can be noticed that *ONDUX* was faster than *U-CRF*, i.e., it needed less than $27s$ to execute the whole process in all extraction tasks, while *U-CRF* needed at least $194s$.

To explain that, it was noticed that in *ONDUX* the Matching step potentially demands the largest amount of time. However, the content-based features used by *ONDUX* are implemented using efficient inverted lists, often used in IR systems. All other steps are linear on the number of terms in the input strings. On the other hand, the extraction process performed by *U-CRF* is slower since the generation of the model for each new extraction task requires verifying several state and transition features for each attribute prior to the proper extraction step.

4.5.5 Comparison with Previous Methods

ONDUX falls in the category of methods that apply learning techniques to extract information from data-rich input strings. As such, it has several points in common with previous methods that have been successfully applied to such a task, such as HMM (Borkar et al. 2001) and CRF (Lafferty et al. 2001). However, it also has unique characteristics that are worth discussing. As CRF is the current state-of-the-art method for this problem, we here compare this method to it. More specifically, we compare *ONDUX* with CRF-based methods in the literature that, ONDUX, rely on previously known data to generate the extraction model. These are the methods presented in Mansuri and Sarawagi (2006) and Zhao et al. (2008), which are referred to as Extended Semi-CRF (ES-CRF) and Unsupervised CRF (*U-CRF*, as in the previous section), respectively.

The first distinction between *ONDUX* and the other two methods is the matching step. This step relies on set of content-based features and does not need to be trained for a specific target source, since it relies only on the known data available on the knowledge base. The main difference between *ONDUX* and the two similar methods, ES-CRF and *U-CRF*, is the way structure-based features, related to positioning and sequencing of attributed values [transition features (Sarawagi 2008)] are learned. In *ONDUX* these features are captured by the PSM model, which, as demonstrated in the experiments, is flexible enough to assimilate and represent variations in the order of the attributes in the input texts and can be learned without user-provided training. *U-CRF* is also capable of automatically learning the order of the attributes, but it cannot handle distinct orderings on the input, since it assumes a single order for the input texts. This makes the application of this method difficult to a range of practical situations.

For instance, in bibliographic data, it is common to have more than one order in a single dataset. Further, the order may vary when taking information from distinct text input sequences, according to the bibliographic style adopted in each input. The order is even more critical in classified ads, where each announcer adopts its own way of describing the object he/she is trying to sell. Another quite common application is to extract data from online shopping sites to store them in a database. The attributes of the offer, such as price, product, discount, and so on, usually appear in a fixed order. In practical applications like these, *ONDUX* is the best alternative method. Further, it is as good as the baselines for any other practical application.

In ES-CRF, distinct orderings are handled, but user-provided training is needed to learn the transition features, similarly to what happens with the standard CRF model, thus increasing the user dependency and the cost to apply the method in several practical situations.

4.6 The *ONDUX* Tool

In order to demonstrate the features of the *ONDUX* method, it created a tool that is called *ONDUX Tool* (Porto et al. 2011). This tool implements all functionalities of the method and it is able to produce all the experimental results reported in Sect. 4.5. Next, this tool is described, it discussed its technical details, and illustrated its main features by means of a case study.

4.6.1 Tool Architecture

Figure 4.3 illustrates the architecture of the *ONDUX* Tool. It consists of three main components: *Data Input*, *ONDUX Engine*, and *Extraction Output*, which are detailed in the following.

The *Data Input* component is responsible for reading and processing two required input files: (1) a structured file containing the occurrences that compose a knowledge base and (2) a text file containing the unstructured records to be extracted.

The knowledge base file must follow a simple XML-based format, which is illustrated in Fig. 4.4. In this figure, each line represents an occurrence that composes the knowledge base. The XML tags correspond to attribute names and the values between the tags correspond to attribute values. In this example, the knowledge base contains occurrences of the attributes Name, Street, City, and Phone.

Besides the tasks of reading and processing the input files, the *Data Input* component builds data structures necessary to the execution of the *ONDUX* method. In particular, as depicted in Fig. 4.3, an inverted index is built for processing the knowledge base.

The inverted index stores important information about the occurrences of each attribute. It contains a vocabulary structure that holds the distinct terms available in the knowledge base. Each entry of this vocabulary contains an occurrence list that stores information about the frequency of each term in a given attribute. This structure is crucial for computing the content-based features used by the *ONDUX* method (See Sect. 3.3).

The *ONDUX Engine* component implements the 3 main steps of the *ONDUX* method: the Blocking step, the Matching step, and the Reinforcement step. The extraction process follows the execution sequence illustrated in Fig. 4.3, thus, a given step can be executed only when the previous step is over.

Finally, the *Extraction Output* component is responsible for presenting the extraction result to the user by exporting it into several formats. This component takes the output of the *ONDUX Engine* component and creates views of the extraction result. As Fig. 4.3 illustrates, the extraction results can be exported into different formats: tables, XML, and CSV.

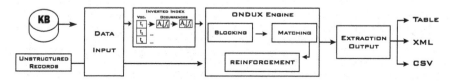

Fig. 4.3 The architecture of the *ONDUX* Tool

Fig. 4.4 Example of a knowledge base file

```
<kb>
    <name> 21st & Century Pools </name>
    <name> Microsoft S.A. </name>
    <street> 630 S Country Rd </street>
    <street> Kennedy Avenue </street>
    <city> New York </city>
    <city> Orlando </city>
    <phone> (516) 447-5242</phone>
    <phone> (55) 92 331-7917</phone>
</kb>
```

4.6.2 Graphical User Interface

In the Tool, the operation of the graphical user interface (GUI) is very intuitive and simple. Figure 4.5 presents a screenshot of the GUI. It includes boxes for loading a file containing the knowledge base and the input file containing unstructured records. The GUI also features buttons for executing each step of the *ONDUX* method, that is, Blocking, Matching, and Reinforcement. Partial results from the extraction process are presented on the screen to the user through tabs.

The Blocking tab presents the blocks resulting from the blocking step. The Matching tab presents the blocks generated in the previous step associated with labels corresponding to attributes, or identified as unmatched. Finally, the Reinforcement tab shows the final extraction result. As illustrated in Fig. 4.5, in this last step, all blocks are associated with an attribute.

An additional tab, PSM, graphically illustrates the positioning and sequencing model (PSM) built for the current extraction process. The last tab, result, presents the extraction result in a tabular format. Finally, the XML and CSV buttons allow the user to export the extraction result in these formats.

4.6.3 Case Study

In this section, it presented a case study in which we use the ONDUX Tool to perform an extraction process over the CORA dataset. As stated in Sect. 4.5.1, CORA is a public dataset that contains unstructured bibliographic references. These references contain several attributes values like: author names, publication titles, page numbers, etc.

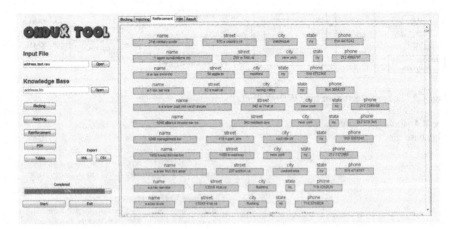

Fig. 4.5 A Screen shot of the *ONDUX* Tool

(a)

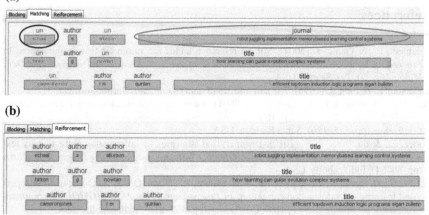

(b)

Fig. 4.6 Matching (**a**) and reinforcement (**b**) steps in the *ONDUX* Tool

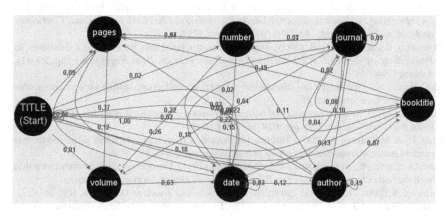

Fig. 4.7 Graphical illustration of the Positioning and Sequencing Model (PSM)

Figure 4.6a, b presents screen shots of the GUI when executing this extraction process. Figure 4.6a shows the result of the Matching step, where almost all blocks were associated with an attribute. The figure also shows cases of blocks that were wrongly labeled and blocks that received the label "un," meaning that these blocks were left unmatched.

The result of the Reinforcement step is depicted in Fig. 4.6b. Now, all blocks are associated with an attribute (i.e., there is no unmatched blocks), and, as illustrated, blocks that were wrongly labeled in the Matching step are now correctly labeled.

As explained in Sect. 4.4, the Reinforcement step relies on the Positioning and Sequencing Model (PSM). Figure 4.7 shows a graphical visualization of the PSM generated by the tool for this extraction task. As already mentioned, this visualization is available on the PSM tab in the tool. In the graph shown, each vertex represents an attribute and the edges represent transition probabilities.

References

Agichtein, E., & Ganti, V. (2004). Mining reference tables for automatic text segmentation. In *Proceedings of the ACM SIGKDD International Conference on Knowledge Discovery and Data Mining* (pp. 20–29), Seattle, USA.

Anderson, T., & Finn, J. (1996). *The new statistical analysis of data*. Berlin: Springer.

Borkar, V., Deshmukh, K., & Sarawagi, S. (2001). Automatic segmentation of text into structured records. In *Proceedings of the ACM SIGMOD International Conference on Management of Data Conference* (pp. 175–186), Santa Barbara, USA.

Cortez, E., & da Silva, A. S. (2010). Unsupervised strategies for information extraction by text segmentation. In *Proceedings of the SIGMOD PhD Workshop on Innovative Database Research* (pp. 49–54), Indianapolis, USA.

Cortez, E., da Silva, A., Gonçalves, M., & de Moura, E. (2010). ONDUX: On-demand unsupervised learning for information extraction. In *Proceedings of the ACM SIGMOD International Conference on Management of Data Conference* (pp. 807–818), Indianapolis, USA.

Kaelbling, L. P., Littman, M. L., & Moore, A. P. (1996). Reinforcement learning: A survey. *Journal Artificial Intelligence Research*, 4(1), 237–285.

Lafferty, J., McCallum, A., & Pereira, F. (2001). Conditional random fields: Probabilistic models for segmenting and labeling sequence data. In *Proceedings of the ICML International Conference on Machine Learning* (pp. 282–289), Williamstown, USA.

Mansuri, I. R., & Sarawagi, S. (2006). Integrating unstructured data into relational databases. In *Proceedings of the IEEE ICDE International Conference on Data Engineering* (pp. 29–41), Atlanta, USA.

McCallum, A. (2012). Cora information extraction collection. http://www.cs.umass.edu/~mccallum/data/cora-ie.tar.gz

Muslea, I. (2012). Rise—a repository of online information sources used in information extraction tasks. http://www.isi.edu/info-agents/RISE/index.html

Peng, F., & McCallum, A. (2006). Information extraction from research papers using conditional random fields. *Information Processing and Management*, 42(4), 963–979.

Porto, A., Cortez, E., da Silva, A. S., & de Moura, E. S. (2011). Unsupervised information extraction with the ondux tool. In *Simpsio Brasileiro de Banco de Dados*, Florianpolis, Brasil.

Sarawagi, S. (2008). Information extraction. *Foundations and Trends in Databases*, 1(3), 261–377.

Zhao, C., Mahmud, J., & Ramakrishnan, I. (2008). Exploiting structured reference data for unsupervised text segmentation with conditional random fields. In *Proceedings of the SIAM International Conference on Data Mining* (pp. 420–431), Atlanta, USA.

Chapter 5
JUDIE

Abstract This chapter presents Joint Unsupervised Structure Discovery and Information Extraction (*JUDIE*) a method for addressing the IETS problem. *JUDIE* was
presented in (Cortez et al. 2011). First, it is introduced the scenario to which *JUDIE* is
targeted to, then we go over the proposed solution detailing all the steps that comprise
JUDIE. Finally, an experimental evaluation of *JUDIE* is presented, comparing its
result with different baselines available in the literature.

Keywords Information extraction · Unsupervised approach · Text segmentation ·
Structure discovery · Knowledge bases · Databases

5.1 The *JUDIE* Method

An important limitation in all previous IETS methods proposed in the literature is
that they rely on the user to implicitly provide the likely structures of the records
found on textual sources. This is true even for the most recent methods that apply
some form of unsupervised learning (Agichtein and Ganti 2004; Cortez et al. 2010;
Mansuri and Sarawagi 2006; Zhao et al. 2008). In most cases, the information on
the likely structures is provided in the training phase, by means of sample records
labeled by a user (Mansuri and Sarawagi 2006). The generated model is then able to
extract information from one record at a time, what requires the user to separate each
individual record prior to providing them as input for the extraction process. In other
cases (Cortez et al. 2010; Zhao et al. 2008), although the structure-based features can
be automatically learned from the unlabeled input records, i.e., no explicit training
is required, these records must still be provided one by one.

This requirement implies into several shortcomings for situations in which many
implicit records are available in a single textual document (e.g., a list of references

---

This chapter has previously been published as (Cortez et al. 2011); reprinted with permission.

E. Cortez and A. S. da Silva, *Unsupervised Information Extraction by Text Segmentation*,   53
SpringerBriefs in Computer Science, DOI: 10.1007/978-3-319-02597-1_5,
© The Author(s) 2013

1/2 cup butter 2 eggs 4 cups white sugar 1/2 cup milk 1 1/2 cups applesauce 2 tablespoons dark rum 2 cups all-purpose flour 1/4 cup cocoa powder 2 teaspoons baking soda ground cinnamon 1/8 teaspoon salt 1 cup raisins 6 chopped pecans 1/4 cup dark rum		

Quantity	Unit	Ingredient
1/2	cup	butter
2		eggs
4	cups	white sugar
1/2	cup	milk
1 1/2	cups	applesauce
2	tablespoons	dark rum
2	cups	all-purpose flour
1/4	cup	cocoa powder
2	teaspoons	baking soda
		ground cinnamon
1/8	teaspoon	salt
1	cup	raisins
6		chopped pecans
1/4	cup	dark rum

Fig. 5.1 Chocolate Cake ingredients (*top*) and structured data extracted from it (*bottom*)

in a research article, or products in an inventory list) or a user is not available for separating the records (e.g., an extractor coupled with a crawler or when processing a stream of documents). Although straightforward methods could be applied to simple cases in which the set of attributes is fixed for all records, dealing with semi-structured records such as heterogeneous bibliographic citations, classified ads, etc., is much more complex. In the case of HTML pages, sometimes it is possible to automatically identify record boundaries and, thus, separate records by using heuristics based on the tags and paths inside the page (Buttler et al. 2001; Embley et al. 1999). However, this is not the most common scenario on the Web and other on-line sources of textual documents, such as social networks or RSS messages.

As an example, consider the Chocolate Cake ingredients available in a pure text message illustrated in Fig. 5.1. To provide a proper input to current IETS methods, a user would have to scan the message and manually separate each record containing the specification of an ingredient in the recipe. Notice the cases in which attributes Quantity and Unit are missing in the input message. Automatically processing several of such messages with current IETS methods is unfeasible, even if they come from the same source.

To deal with this scenario, it is presented *JUDIE* (Joint Unsupervised structure Discovery and Information Extraction), a method for IETS that addresses the problem of automatically extracting several implicit records and their attribute values from a single text input. Unlike previous methods in the literature, ours is capable of detecting the structure of each individual record being extracted without any user intervention. The table in Fig. 5.1 illustrates the output of *JUDIE* when the text on the top is given as input.

To uncover the structure of the input records, a novel algorithm, called *Structure Discover (SD)* algorithm is used, which is based on the *HotCycles* algorithm presented in (de Oliveira and da Silva 2006). The SD algorithm works grouping labels into individual records by looking for frequent patterns of label repetitions, or *cycles*, among a given sequence of labels representing attribute values. It is also shown how to integrate this algorithm in the information extraction process. This is accomplished by successive refinement steps that alternate information extraction and structure discovery. Following, it is presented a brief overview of the method.

5.2 Overview

Given an input text with a set of implicit data records in textual format, such as the one illustrated in Fig. 5.1, the first step of *JUDIE* performs an initial labeling of the candidate values identified in this input with attribute names. As at this point there is no information on the structure of the data records, we resort only to content-based features (Sect. 3.3) for this labeling. Thus, this step, called *Structure-free Labeling*, generates a sequence of labels in which some candidate values may be missing or have received a wrong label. Despite being imprecise, this sequence of labels is accurate enough to allow the generation of an approximate description of the structure of the records in the input text (as demonstrated in the experiments). This is accomplished in the second step of *JUDIE*, called *Structure Sketching*, by using the SD algorithm.

The output of *Structure Sketching* step is a set of labeled values grouped into records that already bear a structure close to the correct one. Thus, from these records it is possible to learn structure-based features (Sect. 3.4). These features can now be used to revise the *Structure-free Labeling* from the first step. This *Structure-aware Labeling* is the third step of the *JUDIE* method. As demonstrated by the experiments, the results produced by this step are more precise than those obtained by the *Structure-free Labeling*, since now content-based and structure-based features are taken into consideration.

JUDIE then takes advantage of the more precise sequence of labels to revise the structure of the records. This new sequence is given as input to the SD algorithm. This is the fourth and final step of the method. It is called *Structure Refinement*. Notice that all of these steps are completely unsupervised.

In what follows it is described in details *JUDIE* by describing the main four steps that comprise it. For that, it is used a running example illustrated in Figs. 5.2a–f. It is considered that the unstructured sequence of tokens corresponding to a list of items of a chocolate cake recipe, shown in Fig. 5.2a, is given as input. The presented method then carries out the task of simultaneously extracting the components of each item, i.e., Quantity (Q), Unit (U) and Ingredient (I), and structuring them into records. The final output is illustrated in Fig. 5.2f.

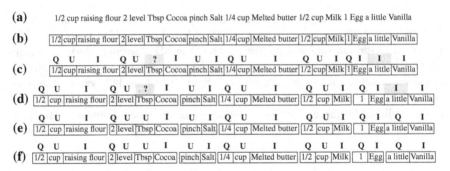

Fig. 5.2 Running example with illustrations of the main steps that comprise *JUDIE*

5.3 Structure-Free Labeling

Given an unstructured input text containing a set of implicit data records in textual
format, such as the one illustrated in Fig. 5.2a, the first step of *JUDIE* consists of
initially labeling potential values identified in this input with attribute names. As at
this point there is no information on the structure of the data records, we resort only to
content-based features (Sect. 3.3) for this labeling. Thus, this step is called *Structure-
free Labeling*. All content-based features used can be computed from a pre-existing
dataset. This pre-existing datasets is called Knowledge Bases. The Knowledge Base
concept is described in Sect. 3.2.

The content-based features considered by *JUDIE* are the ones that were previously
introduced in Sect. 3.3. These features are: (1) Attribute Vocabulary, for exploiting
the common vocabulary often shared by values of textual attributes (2) Attribute
Value Range, for dealing with numeric attributes using the average and the standard
deviation of the values of numeric attributes available on the knowledge base and (3)
Attribute Value Format, for exploiting common writing style often used to represent
values of attributes. In the following, it is described how these features are used to
process the structure-free labeling.

5.3.1 Processing the Structure-Free Labeling

The targets of the structure-free labeling are sequences of tokens in the input text that
are likely to represent attribute values. They are called *candidate values* and they are
defined as follows.

Let $I = t_1, t_2, \ldots, t_n$ be the set of tokens occurring in an input text, such that
no token contains white space. Consider a knowledge base representing attributes
$A_1, \ldots, A_m$. A *likely value* in I is the largest sequence of tokens $s = t_i, t_{i+1}, \ldots, t_{i+k}$
($1 \leq i \leq n, k \geq 0$) from I that occurs as a value, or part of a value, of some attribute
A_j. In the input text I, all likely values and all individual tokens that do not belong
to any likely values are called *candidate values*.

Figure 5.2b illustrates the candidate values found in the input text of Fig. 5.2a. Notice that candidate values such as "raising flour" and "Melted butter" can only be likely values. In the knowledge base used in the experiments for the Cooking Recipes domain, values such as "Milk" and "Salt" are represented. Thus, the corresponding candidate values, in spite of being formed by a single token, are also likely values. On the other hand, "Tbsp" is not present in that knowledge base. Thus, it is an isolated token taken as a candidate value.

Given a candidate value s, the decision on what label must be assigned to it takes into account different domain-dependent features g^k evaluated by feature functions of the form $g^k(s, A)$. To combine these features, it is assumed that they represent the probability of the candidate value s to occur as a value of the attribute A domain, according to the knowledge base. These content-based features are combined using a Bayesian disjunctive operator or, as described in Sect. 3.5.1.

The results of applying the structure-free labeling over the input sequence of Fig. 5.2a is illustrated in Fig. 5.2c, in which capital letters represent labels assigned to candidate values, each label representing an attribute as follows: Q for Quantity, U for Unit and I for Ingredient. Notice that one of the candidate values is marked with a "?", meaning that no label could be assigned to it. This exemplifies one of the anticipated limitations of the structure-free labeling, which is discussed below.

5.3.2 Limitations of the Structure-Free Labeling

The use of very effective domain-dependent features yields a highly precise label assignment in the structure-free labeling step. This claim is supported by the results of extensive experiments reported in this book, involving more than 30 distinct attributes on five distinct datasets.

In spite of that, using such features may represent a problem in two important cases: (1) two (or more) attributes in the same knowledge base are similar with respect to the property being evaluated by the feature function; (2) the property being evaluated is under-represented within the known values of some attribute in the knowledge base. In the first case, wrong labels can be assigned to some segment, i.e., a *label misassignment* occurs. In the second case, there is no support for "safely" assigning a label to that segment, i.e., a *label fault* occurs.

In Fig. 5.2a it is exemplified these two cases by shadowing the labels assigned to two of the candidate values. For the candidate value "Tbsp", the "?" indicates a label fault, while for the candidate value "a little" the shadowed "I" indicates a label misassignment. In this second case, the correct label would be "Q".

For dealing with such cases, state-of-the-art information extraction methods rely on features that also consider the context in which the segment being evaluated occurs within the input text. These features are derived from the structure of the record used as training data (Agichtein and Ganti 2004; Borkar et al. 2001; Cohen and Sarawagi 2004; Cortez et al. 2010; Lafferty et al. 2001; Mansuri and Sarawagi 2006; Zhao et al. 2008).

In this case, it is not possible to use these structure-based features simply because the input text bears no structure. However, imprecise as is, this sequence of labels generated by the structure-free labeling is accurate enough to allow the generation of an approximate description of the structure of the records in the input text. This is accomplished by the second step of the method, called *Structure Sketching*, which is described next.

5.4 Structure Sketching

The goal of the structure sketching step is to organize the labeled candidate values into records, effectively inducing a structure on the unstructured text input. As this step takes as input the labels generated in the structure-free labeling step, in which imprecisions are expected, this structure is considered as a first approximation. The output of this step is a set of labeled values grouped into records that already bear a structure close to the correct one. In this method, this step plays an important role: with the structure of the input text uncovered, we can evaluate structural features and improve the initial labeling from the first step.

The structure sketching step uses a novel algorithm called *Structure Discover (SD)*. Let $\ell_1, \ell_2, \ldots, \ell_n$ be a sequence of labels generated by the structure-free labeling step, in which each label was assigned to a candidate value. The SD algorithm is used to identify in this sequence common subsequences of labels that are frequently repeated in the input text, which are called *cycles*. When a cycle that covers all the input text is found, it can be used to group labels in sub-sequences according to it. Each of these subsequences corresponds to a record grouping values from distinct attributes. These subsequences are called *candidate records*. A detailed discussion of the SD algorithm is postponed to Sect. 5.7.

The result of applying the SD algorithm on the labeled sequence of Fig. 5.2c is shown in Fig. 5.2d. Notice that now candidate values are grouped into distinct sub-sequences, that is, into candidate records. In this example, the cycle found is a simple sequence of the attributes Quantity, Unit and Ingredient.

As this example illustrates, the algorithm is able to deal with irregularities in the candidate records, such as missing or repeated attribute values. Dealing with irregularities is important not only to address natural irregularities often found in real cases, but also to make the process robust to errors caused by the labeling process. In this particular example, while a candidate value of attribute Quantity is indeed missing in the third candidate record, the sequence of three candidate values for attribute Ingredient in the last candidate record is caused by an error in the structure-free labeling step.

As the experimental results indicate, in spite of these and others irregularities (e.g., candidate records with distinct orderings of attribute occurrence), the SD algorithm is able to discover a plausible structure for the input sequence of labels. Again, we refer the reader to Sect. 5.7 for details on the SD algorithm.

With a plausible structure already uncovered by the SD algorithm, it is now possible to compute structure-based features in conjunction with content-based features to improve the initial labeling of the candidate values. This procedure is explained next.

5.5 Structure-Aware Labeling

Consider a candidate record $R = s_1, \ldots, s_r$, where each $s_i (1 \leq i \leq r)$ is a candidate value. Also, consider an attribute A and let ℓ_A be a label used for this attribute. Then, for any candidate value s_i, it can be computed the value of a feature function $f^k(s_i, A, R)$, which is related to the structure of R.

Differently from the content-based features used so far, which are only domain-dependent, structure-based features such as f^k depend on the particular organization of the candidate values within the input text. This means that these features are *source-dependent*.

Like other information extraction methods (e.g., (Cortez et al. 2010; Mansuri and Sarawagi 2006; Zhao et al. 2008), *JUDIE* uses two structure-based features. The first considers the absolute position of the segment and the second considers its relative position, i.e., its occurrence between segment s_{i-1} (if any, i.e., when $i > 0$) and segment s_{i+1} (if any, i.e., when $i < r$).

For computing such features, it is common to use a graph model that represents the likelihood of attribute transitions within the input text (or any other input text from the same source). *JUDIE* uses a probabilistic HMM-like graph model called PSM (Positioning and Sequencing Model), which is described in details in Sect. 3.4. With the structure-based features in hand, they can be used to improve the initial structure-free labeling, as described next.

Given a candidate value s, the decision on which label to assign to it can now consider the structure-based features in addition to the content-based features. As these features are also independent from the content-based ones, since they depend on the source, we again resort to the Bayesian Noisy-OR-Gate (Pearl and Shafer 1988) to combine all features. The process of combining such features is described in Sect. 3.5.2.

The result of applying the structure-free labeling over the candidate records of Fig. 5.2d is illustrated in Fig. 5.2e. Notice that with the addition of the structure-based features, the candidate value "Tbsp" is now correctly labeled as U for Unit (this term is indeed used in place of "tablespoon"). For the same reason, candidate value "a little" is now correctly labeled as Q for Quantity.

As this example suggests, in general, combining structure-based and content-based features produce more precise results than the initial structure-free labeling. This trend is clearly indicated by the experiments.

JUDIE then takes advantage of this more precise sequence of labels to also revise the structure of the records. This new sequence is given as input to the SD algorithm. This is the fourth and final step of the method.

5.6 Structure Refinement

This last step of the method simply consists in applying again the SD algorithm. This time, however, it takes as input the labels generated by the structure-aware labeling. As the labeling produced by this step is more precise, the result is a more accurate structure. This is also indicated by the experimental results.

To illustrate it, notice that in Fig. 5.2f the last candidate record from Fig. 5.2g has now been split in two different records by the SD algorithm. In the next section it is described in details the the SD algorithm.

5.7 The SD Algorithm

The main intuition behind the SD algorithm is that it is possible to identify patterns of sequences by looking for cycles into a graph that models the ordering of labels in the labeled input text. This graph, called *Adjacency Graph*, is defined below.

Adjacency Graph. Consider the sequence $s_1, s_2, \ldots, s_n$ of candidate values in the input text, such that s_i is labeled with ℓ_i. The ordered list $L = \langle \ell_1, \ell_2, \ldots, \ell_n \rangle$ is called an *Adjacency List*. An *Adjacency Graph* is a digraph $G = \langle V, E \rangle$ in which V is the set of all distinct labels in L, plus two special labels *begin* and *end*, and E is the set of all pairs $\langle \ell_i, \ell_j \rangle$ in E for all i, j such that $j = i + 1$ ($1 \leq i \leq n - 1$), plus two special edges $\langle begin, \ell_1 \rangle$ and $\langle \ell_n, end \rangle$.

Figure 5.3 illustrates portions of an Adjacency List built from a sample unstructured text containing a number of implicit bibliographic data records. This sample is a simplified version of a real bibliographic data source such as *CORA* (a dataset used in the experiments) represented by some of the attributes involved (e.g., no volume or page information is represented). This sample, however, exemplifies some of the problems faced when processing real textual inputs.

Figure 5.3 also illustrates an Adjacency Graph built from this Adjacency List. In this graph, nodes corresponding to attributes are represented by ellipses identified by their respective labels in the adjacency list. Nodes *begin* and *end* are considered as if they occurred only once in this list, respectively, before and after the sequence of

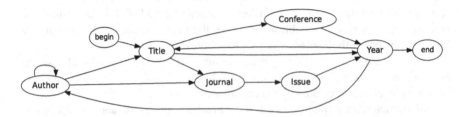

Fig. 5.3 An adjacency list and an adjacency graph for an input text with bibliographic data

candidate values $s_1, \ldots, s_n$. Their role is simply to serve as references for the graph processing algorithms used by *JUDIE*.

The two long paths ⟨Author,Title,Conference,Year⟩ and ⟨Author, Title,Journal, Issue,Year⟩ correspond, respectively, to publications in conferences and journals. Notice, however, that some edges indicate the occurrence of implicit records with missing attributes. This is the case of the edge ⟨Author,Journal⟩ that indicates a missing value for Title. Also notice that ⟨Year,Author⟩ and ⟨Year,Title⟩ intuitively indicate records ending with an Year candidate value leading to another record that may begin with either Author or Title. Indeed, the first implicit record in the input text begins with a Title candidate value.

The occurrence of implicit records with missing attributes is a very common issue in most real cases. This situation occurs either due to errors in the labeling process, specially in the case of the structure-free labeling, or because the implicit record indeed has no value for some attributes.

An important aspect the SD algorithm exploits in the adjacency graph is the occurrence of cycles. A cycle is a sequence of adjacent nodes ⟨$\ell_i, \ldots, \ell_{i+k}, \ell_i$⟩. For convenience, the notation [$\ell_i, \ldots, \ell_{i+k}$] is used, omitting the last node, which is always equal to the first one.

The different configurations of implicit records, i.e., the set of attributes composing them and the order in which their candidate values appear, can be detected by looking for cycles in the adjacency graph. This is the case of cycles [Author,Title,Conference, Year], [Author,Title,Journal,Issue,Year] and [Title,Conference,Year] in Fig. 5.3.

Two important issues arise when using the adjacency graph to analyze the possible record structures in the input text: (1) in which order the labels in the cycle occur in the input text and (2) which cycles correspond to actual implicit records in the input text. To deal with both issues, it is verified the correspondence between cycles and the sequence of labels in the adjacency list. For the definitions below, let G be an adjacency graph generated from an adjacency list L.

Coincident Cycles. Two cycles c_a and c_b are said to be *coincident*, meaning that they represent the same cycle in G, if they include the same edges in the same order, but beginning and ending at a different node in the cycle.

Cycle Instances and Viable Cycles. Let $c = [\ell_i, \ldots, \ell_{i+k}]$ be a cycle in G. Any sequence $\ell_i, \ldots, \ell_{i+k}$ in L is said to be an *instance* of c. The cycle c is said to be *viable* if there is at least one instance of c in L.

Dominant Cycles. Let $\{c_1, \ldots, c_n\}$ be a set of coincident cycles. The viable cycle c_i for which the order of labels is the most frequent in L is called the *dominant cycle*.

To exemplify these concepts, cycles c_a = [Author,Title,Conference,Year] and c_b = [Title,Conference,Year,Author] are coincident in the adjacency graph of Fig. 5.3. By looking into the adjacency list, it can be found that c_b is the dominant cycle.

These concepts are used by the SD algorithm (Algorithm 2). This algorithm works by first identifying all dominant cycles in the adjacency graph and then processing each of these cycles in the order of their sizes, the largest cycles being processed

---

Algorithm 2 Structure Discovery Algorithm

1: $L \leftarrow adjlist(I)$;
2: $G \leftarrow adjgraph(L)$;
3: **for all** *single cyle* $[\ell, \ell]$ *in* G **do**
4:    Replace all sequences $\ell, \ldots, \ell$ by one single element $\ell+$ in L
5: **end for**
6: $G \leftarrow adjgraph(L)$;
7: $C \leftarrow dominant_c ycles(G)$;
8: $i \leftarrow 0$;
9: **while** $C \neq \emptyset$ **do**
10:    $dc_i \leftarrow next(C)$;
11:    **for** each instance $\ell_1, \ldots, \ell_k$ of dc_i in L **do**
12:       Replace $\ell_1, \ldots, \ell_k$ by r_i in L;
13:    **end for**
14:    $i + +$;
15: **end while**

---

first. Notice that nodes *begin* and *end* never participate in any cycle, since they are both connected to the graph by a single edge.

In Lines 2 and 3, the Adjacency List and the Adjacency Graph are created. Next, in Lines 4 and 5, the algorithm detects all single cycles in the graph in order to remove all sequences of a same label from the adjacency list. Such sequences usually represent multivalued attributes (e.g., lists) that must be considered as a single component in the records being identified. Thus, these sequences are replaced by a single label $\ell+$ in the adjacency list.

In Line 6, a new Adjacency Graph is generated for reflecting the removal of theses sequences. If we consider the graph in Fig. 5.3, the only effect will be the removal of the cycle involving Author.

In Line 7, the algorithm extracts all dominant cycles from G. Next, these dominant cycles are used to structure their instances in the input text. This is carried out by the loop in Lines 9 to 13. In Line 10, the function *next* selects and removes the largest dominant cycles from C and, in Lines 11 and 12, the instances of the cycles in the adjacency list L are replaced by an indication that a record has been formed with each of these instances. Thus, in this algorithm, records are taken as cycle instances whose boundaries are determined by matching cycles derived from the graph to the adjacency list (Line 11).

It can be noticed the importance of processing larger cycles first. Considering the graph in Fig. 5.3, if the cycle $c_b = $ [Title, Conference, Year] was processed before $c_a = $ [Author, Title, Conference, Year], part of each instance of c_b would be taken as a instance of c_a. This process continues while there are cycles unprocessed in C.

For the Adjacency List and the Adjacency Graph of Fig. 5.3, the sequence of dominant cycles that would be processed is the following: [Author, Title, Journal, Issue, Year], [Author, Title, Conference, Year], [Author, Journal, Issue, Year], [Title, Conference, Year] and [Title, Year].

5.8 Experimental Evaluation

In this section, it is described the experiments that were performed to evaluate *JUDIE* using five distinct datasets. First, it is described the experimental setup used to assess *JUDIE*'s performance. Then, it is reported the quality of the extraction results for each dataset.

5.8.1 Setup

The datasets employed in the experiments and the data sources used to generate the knowledge bases for *JUDIE* are summarized in Table 5.1. We notice that some of these datasets are the same employed in the evaluation of other information extraction methods. We also recall that *JUDIE* takes as input sets of records without any explicit delimiters between them, as illustrated in Fig. 5.1.

The dataset of the *Cooking Recipes* domain was previously used in (Barbosa and Freire 2010). In order to build the knowledge base for this domain, we have collected structured recipes from FreeBase.[1] For the *Product Offers* domain, the dataset is formed by unstructured strings containing lists of product offers from 25 Brazilian e-commerce stores. Data for building the respective knowledge base has been taken from Neemu,[2] a Brazilian price comparison website. For the *Postal Addresses* domain, both the dataset and the data source used to build the knowledge base have been obtained from *Bigbook*, a dataset available in the RISE repository[3] and that has been previously used in (Zhao et al. 2008 and Cortez et al. 2010).

For the *Bibliography* domain, the dataset is part of the Cora Collection[4] and is composed of a large diversity of bibliographic citations in distinct styles and formats. It includes citations to journal articles, conference papers, books, technical reports, etc. The data source for building the knowledge base, PersonalBib, is also a dataset of bibliographic citations that has been used in (Mansuri and Sarawagi 2006). Finally, for the *Classified Ads* domain we have taken the dataset previously used in (Cortez et al. 2010). This dataset is composed of unstructured strings containing ads from Brazilian newspaper websites. For building the knowledge base, data from a database available was collected from the website of a major Brazilian newspaper.

For all performed experiments, the extraction results were evaluated for each individual attribute (attribute-level) and for each record type as whole (record-level). As evaluation metrics, the well known precision, recall and F-measure were used as defined next.

Let B_i be a reference set and S_i be a test set to be compared with B_i. We define precision (P_i), recall (R_i) and F-measure (F_i) respectively as:

---

[1] http://www.freebase.com

[2] http://www.neemu.com

[3] http://www.isi.edu/info-agents/RISE

[4] http://www.cs.umass.edu/~mccallum/data

Table 5.1 Domains, datasets and KB data sources used in the experiments

Domain	Dataset	Text inputs	Attributes	Source	Attributes	Records
Cooking recipes	Recipes	500	3	FreeBase	3	100
Product Offers	Products	10000	3	Neemu.com	3	5000
Postal addresses	BigBook	2000	5	BigBook	5	2000
Bibliography	CORA	500	3–7	PersonalBib	7	395
Classified Ads	WebAds	500	5–18	Folha On-line	18	125

$$P_i = \frac{|B_i \cap S_i|}{|S_i|}, \; R_i = \frac{|B_i \cap S_i|}{|B_i|} \text{ and } F_i = \frac{2(R_i.P_i)}{(R_i + P_i)} \qquad (5.1)$$

In order to present attribute-level results, we calculate precision, recall and F-measure according to the above equations by considering B_i as the set of terms that compose the values of a given attribute a_i and S_i the set of terms assigned to a_i by *JUDIE*. Likewise, for record-level results, we calculate precision, recall and F-measure by considering each record set B_i as the set of field values in a given structured record C_i and S_i the set of field values extracted for C_i by *JUDIE*.

5.8.2 General Quality Results

In this section, it is analyzed the general quality of the extraction task performed by *JUDIE* on the datasets described in Table 5.1. For each domain, the extraction task was run five times, each time selecting different data samples for the data extraction task and for building the respective knowledge bases. For all performed extractions, it is reported the average F-measure obtained for all runs. It can also be noticed that there is no intersection between the knowledge bases and the corresponding datasets used in the experiments.

Tables 5.2a–c and 5.3a–b present attribute-level F-measure values that assess the extraction quality in each dataset. Column "C1" refers to results obtained after the Structure-free Labeling and Structure Sketching steps, which correspond to what is called Phase 1, and Column "C2" refers to results obtained after the Structure-aware Labeling and Structure Discovery steps, which correspond to what is called Phase 2. Column "G" presents the gain achieved from Phase 1 to Phase 2.

Each of these columns assesses a distinct aspect of the method. Results in column "C1" assess how well the content-based source-independent features alone have been able to assign correct labels to the input text, while results in column "C2" also account for the use of structure-based source-dependent features learned from the input text itself.

To provide a perspective on the contribution of each feature to the overall extraction quality, it is also presented F-measure values obtained when each type of feature is individually used. The cases considered are: (1) either the *fitness* function

Table 5.2 Attribute-level results for datasets Recipes, Products and BigBook

Attribute	Phase 1			Phase 2		
	FI/NM	FO	C1	S+P	C2	G %
(a) Recipes						
Quantity	0.81	0.69	0.89	0.78	0.96	7.1
Unit	0.86	0.46	0.91	0.82	0.94	3.9
Ingredient	0.84	0.74	0.91	0.76	0.96	4.9
Average	0.84	0.63	0.90	0.79	0.95	5.3
(b) Products						
Name	0.77	0.37	0.85	0.69	0.90	5.3
Brand	0.74	0.52	0.83	0.71	0.92	10.5
Price	0.89	0.92	0.93	0.88	0.95	1.9
Average	0.80	0.60	0.87	0.76	0.92	5.8
(c) BigBook						
Name	0.79	0.48	0.94	0.63	0.97	2.6
Street	0.82	0.40	0.95	0.75	0.97	2.6
City	0.92	0.39	0.94	0.84	0.97	2.8
State	0.89	0.63	0.96	0.88	0.97	1.3
Phone	0.94	0.93	0.95	0.89	0.97	2.3
Average	0.87	0.57	0.95	0.80	0.97	2.3

for textual attributes (Eq. 3.2) or the *NM* function for numeric attributes (Eq. 3.3) is used (Column "FI/NM"); (2) only the *format* function (Eq. 3.5) is used (Column "FO"); and (3) only the *pos* and *seq* (Eq. 3.8) functions are used (Column "S+P"). Recall that "C1" results are obtained by combining in Phase 1 functions *fitness* (or *NM*) and *format* by using Eq. 3.10, and that "C2" results are obtained by combining in Phase 2 functions *fitness* (or *NM*), *format*, *pos* and *seq* by using Eq. 3.11.

As anticipated, it can be observed that the attribute-level results obtained in Phase 1 by combining features are already acceptable and, more importantly, are sufficient to yield a reasonable approximation of the records' structure. Furthermore, the *fitness* and *MN* functions are, in general, more accurate than the *format* function. However, their combination, as proposed in *JUDIE*, leads to better results in all cases.

Phase 2 results are higher in all cases. While in most cases the gain is under 6 %, there are interesting cases in which this gain is above 10 %. For example, Title and Journal are attributes that present a large content overlap in the Bibliography dataset. Due to this problem, the percentage of labels incorrectly assigned to values of Title and Journal in Phase 1 was 25 and 16 % respectively. In Phase 2, the majority of these misassignments were corrected. A large gain was also observed in the case of attribute Brand from the Products dataset. Because brand names are formed by terms usually not available in the knowledge base, more than 12 % of values of this attribute were left unmatched in Phase 1. The structure-based features used in Phase 2 helped to recover these errors.

The behavior of the method when dealing with numeric attributes deserves specific comments. Considering the eight numeric attributes from the five distinct datasets used in the experiments, the *NM* function yielded an average attribute-level

Table 5.3 Attribute-level results for datasets CORA and WebAds

Attribute	Phase 1			Phase 2		
	FI/NM	FO	C1	S+P	C2	G %
(a) CORA						
Author	0.79	0.60	0.83	0.65	0.88	5.9
Title	0.60	0.52	0.70	0.48	0.79	13.8
Booktitle	0.82	0.46	0.81	0.67	0.86	6.2
Journal	0.69	0.53	0.72	0.62	0.84	16.9
Volume	0.84	0.88	0.88	0.72	0.90	2.9
Pages	0.79	0.80	0.83	0.73	0.86	3.9
Date	0.72	0.76	0.79	0.69	0.87	9.5
Average	0.75	0.65	0.79	0.65	0.86	8.1
(b) WebAds						
Bedroom	0.75	0.36	0.79	0.48	0.82	3.8
Living	0.81	0.46	0.85	0.69	0.89	5.6
Phone	0.79	0.84	0.80	0.62	0.87	8.8
Price	0.85	0.85	0.86	0.66	0.92	7.2
Kitchen	0.80	0.29	0.79	0.73	0.83	4.9
Bathroom	0.73	0.59	0.75	0.69	0.77	2.9
Suite	0.85	0.45	0.87	0.60	0.89	2.4
Pantry	0.79	0.50	0.77	0.66	0.80	3.7
Garage	0.78	0.52	0.79	0.73	0.84	6.6
Pool	0.77	0.63	0.78	0.78	0.82	5.2
Others	0.70	0.44	0.72	0.68	0.73	1.6
Average	0.78	0.54	0.80	0.67	0.84	4.8

F-measure of 0.83 when used alone. This result is close to that obtained for the non-numeric attributes using the *fitness* function. For instance, with phone numbers, using only the *NM* function *JUDIE* obtained F-measure values of 0.94 and 0.79 for the BigBook and WebAds datasets, respectively. Moreover, the *format* function is also used with these attributes, but, in this case, unlikely to what happens with the *NM* function, values are not normalized. When used alone, this function also yielded an average F-measure of 0.83. Finally, the structure-based features helped to improve these results. When combined with the other two features, an average F-measure of 0.91 was obtained. As it can be seen, *JUDIE* achieves equally good results with both numeric and textual attributes.

Table 5.4 presents, for each dataset, record-level F-measure results obtained in Phase 1 and Phase 2. While results in Phase 1 are also acceptable (most of them above 0.7), improvements in labeling achieved in Phase 2 had a very positive effect. Indeed, in Phase 2 record-level F-measure has achieved results above 0.8 for four out of five datasets and, in all cases, gains have been above 7 %. Notice, for instance, the case of the CORA dataset, in which the gain is higher than 19 %, reflecting the improvements obtained by the structure-aware labeling step. As it can be noticed, adding the structure-based features (only possible in Phase 2) also leads to significant improvements regarding record-level results.

Table 5.4 General record-level results for each dataset

Dataset	Phase 1	Phase 2	Gain %
Recipes	0.79	0.90	13.2
Products	0.82	0.88	7.2
BigBook	0.86	0.93	8.8
CORA	0.69	0.83	19.3
WebAds	0.70	0.77	9.7

5.8.3 Impact of the Knowledge Base

In (Cortez et al. 2010) the authors present an experiment to evaluate how dependent on the composition of the knowledge base is the quality of the extraction results.

In the case of *JUDIE* such study is even more important for the following reasons: (1) the extraction process entirely relies on the initial Structure-free Labeling step, which is solely based on content-based features learned from the knowledge base; (2) while in the closest competitor, ONDUX (Cortez et al. 2010), the knowledge base is used only for matching, *JUDIE* also deploys a format feature based on its values. Thus, in *JUDIE* the knowledge base plays a crucial role, as it is shown in this experimental evaluation.

Here we compare *JUDIE* with ONDUX and U-CRF. These two methods are the current state-of-the-art unsupervised IETS methods. U-CRF was developed by adapting the publicly available implementation of CRF by Sunita Sarawagi[5] according to (Zhao et al. 2008) and using additional features described in (Lafferty et al. 2001) (e.g., dictionary features, word score functions, transition features, etc.). As required by U-CRF, a batch of input strings is used to infer the order of the attribute values. Based on the information provided in (Zhao et al. 2008), this batch is built using a sample of 10 % of these strings.

As in (Cortez et al. 2010), this experiment was performed using the *BigBook* dataset from the RISE repository. The knowledge base for ONDUX and JUDIE and the reference table for U-CRF were built by using sets of records already extracted. Once again, it can be noticed that there is no intersection between these records and the corresponding datasets used in this experiment. Recall that while ONDUX and U-CRF received the input in a record-by-record basis, *JUDIE* received a single input text containing all 2000 records with no explicit delimiters between them.

The experiment consisted of varying the number of known terms common to the knowledge base (or reference table in the case of U-CRF) and the input test records from 50 to 1000 terms and evaluating the extraction quality in terms of average attribute-level F-measure. The results are presented in Fig. 5.4a.

The first important observation regarding this graph is that *JUDIE* is, as expected, more dependent on the knowledge base than ONDUX and U-CRF. Indeed, only when the number of shared terms approaches 1000, it reaches the same quality level as the

---

[5] http://crf.sourceforge.net/

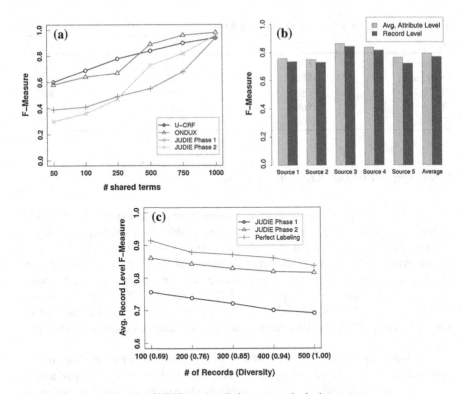

Fig. 5.4 Results obtained by *JUDIE* varying distinct aspects in the input texts

baselines. This occurs because in both ONDUX and U-CRF the structure-based and content-based features are independent, while in *JUDIE*, as previously explained, content-based features are used to induce structured-based features through successive refinement steps.

Indeed, if content-based features are not enough, the induction of structure-based features fails. This can be observed in Fig. 5.4a, where the attribute-level F-measure values obtained with less than 250 common terms are very low. For this level of term intersection, the results of *JUDIE*'s Phase 1, i.e., before any refinement, are better than the results of its Phase 2, in which structure-based features are also considered.

In spite of this limitation, *JUDIE* achieves results comparable to the state-of-the-art baselines for a task considerably harder, that is, extracting information while simultaneously uncovering its underlying structure. As already explained, this underlying structure is assumed as provided in the baseline methods. In Sect. 4.5.5, it is presented a detailed comparison between *JUDIE* and these baselines using other datasets.

5.8.4 Impact of Structure Diversity

In this section we study how *JUDIE* deals with different types of structure observed in the implicit records found in the input text. For this we consider two different scenarios, structure diversity in different sources and within a single source. These two scenarios are discussed in the following.

5.8.4.1 Structure Diversity in Different Sources

To discuss the first scenario we use the Classified Ads domain, for which the knowledge base was build using data from one source and the input texts came from other five distinct sources. In the experiments reported below, each source corresponds to a different input text. Here, our goal is to demonstrate that the content-based features learned from data taken from one source can be used to induce the structure-based features for several related input texts from other distinct sources in the same domain.

In Fig. 5.4b it is shown the attribute-level and record-level F-measure values obtained for each different source given as input to *JUDIE*. In all cases, the values are above 0.7 and for two cases they are above 0.8. This indicates that *JUDIE* is source-independent, since it was able to correctly uncover the structure of implicit records in each source while also achieving good extraction level quality. This occurs despite the differences in structure of the implicit records in each source.

5.8.4.2 Structure Diversity Within a Single Source

For discussing the second scenario we use the Bibliography domain in which the knowledge base was built from the PersonalBib dataset (Mansuri and Sarawagi 2006) and single input texts came from the CORA collection. In this case we aim at showing how *JUDIE* deals with a heterogeneous dataset in terms of structure.

By examining the distribution of citation styles among the 500 implicit records available in the CORA dataset, a total of 33 distinct styles were identified, but only six of them account for more than 90 % of the citations.[6]

For these experiments, it was generated different input texts containing 100–500 implicit records randomly selected from the CORA dataset. We then process each of these input texts separately with *JUDIE* using the knowledge base described above. The process was repeated 10 times for each input text size.

To characterize the diversity of each input text it was used the Shannon Index (Shannon 2001), which is frequently used to measure diversity in categorical datasets. This index is defined as: $H = -\sum_{i=1}^{S} p_i ln(p_i)$, where S is the total number of styles (33 in this case) and p_i is the relative frequency of each style i found in the input text. As the H index does not return values between 0 and 1, we normalize H values obtained for each input text by the maximum possible value for H. This value occurs

---

[6] A citation style characterized by the set of attributes composing the record and their ordering.

when the input contains all citations available in the CORA dataset, that is, when all 33 different citation styles are present in the input text.[7] Thus, the closer the normalized *H* value is to 1, the greater is the diversity of the input text.

The results obtained are presented in Fig. 5.4c in terms of the average record-level F-measure, considering Phases 1 and 2 of *JUDIE*. As a baseline, we use the record-level F-measure we would obtain if the labeling of attribute values was perfect in the input texts. In the X-axis, it is presented the number of records and the diversity of each input text in terms of the normalized *H* index.

This figure shows that the SD algorithm deals very well with structure diversity if the labels are correctly assigned, as it can be seen by comparing the curves representing *JUDIE* Phase 1 and the perfect labeling. As it can also be observed, the improvements on the quality of the labeling provided by adding the structure-based features in Phase 2 impacts positively on the quality of the structure discovery. Indeed, record-level F-measure values obtained in Phase 2 are close to those obtained with the perfect labeling.

5.8.5 Comparison with Previous Work

In this section it is presented a comparison between the results obtained by *JUDIE* with those obtained by two state-of-the-art IETS methods, namely ONDUX (Cortez et al. 2010) and U-CRF (Zhao et al. 2008).

This comparison is made by reproducing in Tables 5.5a–c the attribute-level results obtained for three datasets, which were reported in (Cortez et al. 2010) for the two methods, along with the results we obtain by running *JUDIE* over the same datasets.

Tables 5.5a–c the attribute-level results obtained for three datasets, which were reported in (Cortez et al. 2010) for the two methods, along with the results we obtain by running *JUDIE* over the same datasets.

While ONDUX was first presented and fully described in that paper, U-CRF was used there as a baseline. The details on its implementation are summarized in Sect. 5.8.3. In all cases, it was used the same sources for generating the knowledge bases and the input texts. Recall again that among the three methods, *JUDIE* is the only one that is able to both discover the structure and extract information automatically.

As a general observation, in spite of the fact the *JUDIE* faces a harder task, its performance was very close to that of ONDUX. In most cases, ONDUX outperformed *JUDIE*, but there are a few cases in which *JUDIE* performed better than ONDUX. These cases are explained mainly by the use of the format feature in *JUDIE*. Such feature is not considered in ONDUX.

In comparison with U-CRF, *JUDIE* performed worse on the BigBook dataset, but better on the CORA and WebAds datasets. This was expected, since these datasets are much more irregular in terms of structure than the first one.

---

[7] $H = 2.23671$ for the input containing 500 records.

Table 5.5 Comparison of results

Attribute	*JUDIE*	ONDUX	U-CRF
(a) BigBook			
Name	0.967	0.996 (2.97%)	0.995 (2.86%)
Street	0.970	0.995 (2.5%)	0.993 (2.37%)
City	0.971	0.995 (2.43%)	0.990 (1.92%)
State	0.971	1.000 (2.95%)	0.999 (2.84%)
Phone	0.975	1.000 (2.57%)	0.988 (1.34%)
Average	0.971	0.997 (2.70%)	0.993 (2.27%)
(b) CORA			
Author	0.881	0.922 (4.65%)	0.876 (-0.57%)
Title	0.794	0.792 (-0.25%)	0.694 (-12.59%)
Booktitle	0.855	0.892 (4.33%)	0.560 (-34.50%)
Journal	0.843	0.908 (7.71%)	0.553 (-34.40%)
Volume	0.901	0.958 (6.33%)	0.430 (-52.28%)
Pages	0.861	0.849 (-1.39%)	0.503 (-41.58%)
Date	0.865	0.895 (3.47%)	0.488 (-43.58%)
Average	0.857	0.888 (3.60%)	0.586 (-31.60%)
(c) WebAds			
Bedroom	0.818	0.861 (5.25%)	0.791 (−3.30%)
Living	0.893	0.905 (1.34%)	0.724 (−18.93%)
Phone	0.873	0.926 (6.12%)	0.754 (−13.59%)
Price	0.923	0.936 (1.41%)	0.786 (−14.84%)
Kitchen	0.830	0.849 (2.29%)	0.788 (−5.06%)
Bathroom	0.773	0.792 (2.51%)	0.810 (4.84%)
Suite	0.894	0.881 (−1.50%)	0.900 (0.62%)
Pantry	0.800	0.796 (−0.55%)	0.687 (−14.17%)
Garage	0.844	0.816 (−3.28%)	0.714 (−15.37%)
Pool	0.818	0.780 (−4.66%)	0.683 (−16.52%)
Other	0.732	0.796 (8.68%)	0.719 (−1.84%)
Average	0.836	0.849 (1.52%)	0.760 (−9.16%)

5.8.6 Performance Issues

In Table 5.6 it is presented the running times of the experiments executed with *JUDIE*. For the datasets used to run comparative experiments with the baselines ONDUX and U-CRF, it was also included the running times of these systems. As the number of implicit input records is different for each dataset, it is presented both the total running time and the average running time by record.

Before discussing the results, it can be noticed that *JUDIE* running times depend on two main factors: the number of implicit input records and the diversity in the structure of these records. Regarding the first factor, in all steps the input is scanned once. Thus, there is a linear influence of this factor. As for the second factor, having more diverse records in terms of structure implies that a larger number of edges

Table 5.6 *JUDIE* running times in comparison with baselines

Datasets	Total (secs.)			Avg. per record (msecs.)		
	JUDIE	ONDUX	U-CRF	*JUDIE*	ONDUX	U-CRF
Recipes	37.5	–	–	75.1	–	–
Products	69.2	–	–	6.9	–	–
BigBook	50.2	14.1	297.1	25.1	7.1	148.5
CORA	74.4	10.7	185.9	148.8	21.4	371.8
WebAds	59.2	8.0	2701.9	118.5	16.0	5403.7

will occur in the Adjacency Graph and in the PSM. Thus, processing these graphs has higher costs for more heterogeneous structures. This explain why the average running times per record are higher for CORA and WebAds, which, as discussed in Sect. 5.8.4, are the more diverse datasets in the experiments.

Nevertheless, these running times are in the same order of magnitude as those of ONDUX and are, in general, smaller than those of U-CRF. ONDUX is faster since it executes fewer steps and does not include a structure discovery step. U-CRF has a worst performance due to costly inference steps, particularly when dealing with diverse structures, and due to the use of a larger number of features than ONDUX and *JUDIE*.

References

Agichtein, E., & Ganti, V. (2004). *Mining reference tables for automatic text segmentation. Proceedings of the ACM SIGKDD International Conference on Knowledge Discovery and Data Mining* (pp. 20–29). USA: Seattle.

Barbosa, L., & Freire, J. (2010). *Using Latent-structure to Detect Objects on the Web. Proceedings of the Workshop on Web and Databases* (pp. 1–6). USA: Indianapolis.

Borkar, V., Deshmukh, K., & Sarawagi, S. (2001). *Automatic Segmentation of Text into Structured Records. Proceedings of the ACM SIGMOD International Conference on Management of Data Conference* (pp. 175–186). USA: Santa Barbara.

Buttler, D., Liu, L., & Pu, C. (2001). *A fully automated object extraction system for the World Wide Web. Proceedings of the International Conference on Distributed Computing Systems* (pp. 361–370). USA: Washington.

Cohen, W. W., & Sarawagi, S. (2004). *Exploiting dictionaries in named entity extraction: combining semi-markov extraction processes and data integration methods. Proceedings of the ACM SIGKDD International Conference on Knowledge Discovery and Data Mining* (pp. 89–98). USA: Seattle.

Cortez, E., da Silva, A., Gonçalves, M., & de Moura, E. (2010). *ONDUX: On-Demand Unsupervised Learning for Information Extraction. Proceedings of the ACM SIGMOD International Conference on Management of Data Conference* (pp. 807–818). USA: Indianapolis.

Cortez, E., da Silva, A. S., de Moura, E. S., & Laender, A. H. F. (2011). *Joint unsupervised structure discovery and information extraction. Proceedings of the ACM SIGMOD International Conference on Management of Data Conference* (pp. 541–552). Athens: Greece.

de Oliveira, D. P., & da Silva, A. S. (2006). *Extrao de dados de pginas web ricas em dados na ausncia de informaes estruturais.* Master's thesis: Universidade Federal do Amazonas.

Embley, D., Campbell, D., Jiang, Y., Liddle, S., Lonsdale, D., Ng, Y., et al. (1999). Conceptual-model-based data extraction from multiple-record web pages. *Data and Knowledge Engineering*, *31*(3), 227–251.

Lafferty, J., McCallum, A., & Pereira, F. (2001). *Conditional random fields: Probabilistic models for segmenting and labeling sequence data. Proceedings of the ICML International Conference on Machine Learning* (pp. 282–289). USA: Williamstown.

Mansuri, I. R., & Sarawagi, S. (2006). *Integrating unstructured data into relational databases. Proceedings of the IEEE ICDE International Conference on Data Engineering* (pp. 29–41). USA: Atlanta.

Pearl, J., & Shafer, G. (1988). *Probabilistic reasoning in intelligent systems: Networks of plausible inference*. Morgan Kaufmann.

Shannon, C. E. (2001). A mathematical theory of communication. *ACM SIGMOBILE Mobile Computing and Communications Review*, *5*(1), 3–55.

Zhao, C., Mahmud, J., & Ramakrishnan, I. (2008). *Exploiting structured reference data for unsupervised text segmentation with conditional random fields. Proceedings of the SIAM International Conference on Data Mining* (pp. 420–431). USA: Atlanta.

Chapter 6
iForm

Abstract This chapter presents *iForm*, a method for automatically using data-rich text for filling form-based input interfaces that rely on the presented unsupervised approach to deal with the Information Extraction by Text Segmentation problem. *iForm* was first presented in Toda et al. (2009, 2010). In the following is described the scenario where *iForm* is applied, and the method in detail. A set of experiments is also reported that shows that *iForm* is effective and works well in different scenarios.

Keywords Information extraction · Web form-filling · Text segmentation · Data management · Web · Databases

6.1 The Form-Filling Problem

The Web is abundant in applications where casual users are required to enter data to be stored in databases for further processing. The most common solution in these cases is to design form-based interfaces which contain *multiple data input fields*, such as text boxes, radio buttons, pull-down lists, check boxes, and other input mechanisms. Unlike typical search forms, these Web input forms usually have a larger number of fields. Figure 6.1 presents a real Web form from the cars domain. It can be noticed that, as stated above, this form contains multiple fields.

Although these interfaces are popular and effective, in many cases interfaces that accept *data-rich free text* as input, i.e., documents or text portions that contain implicit data values, would be preferable. Indeed, in many cases the data required to fill the form fields could be taken from text files in which they are already available. For instance, a job applicant may use data taken from a resume text file to fill several fields of forms in many different job search sites.

This chapter has previously been published as (Toda et al. 2010); reprinted with permission.

E. Cortez and A. S. da Silva, *Unsupervised Information Extraction by Text Segmentation*,   75
SpringerBriefs in Computer Science, DOI: 10.1007/978-3-319-02597-1_6,
© The Author(s) 2013

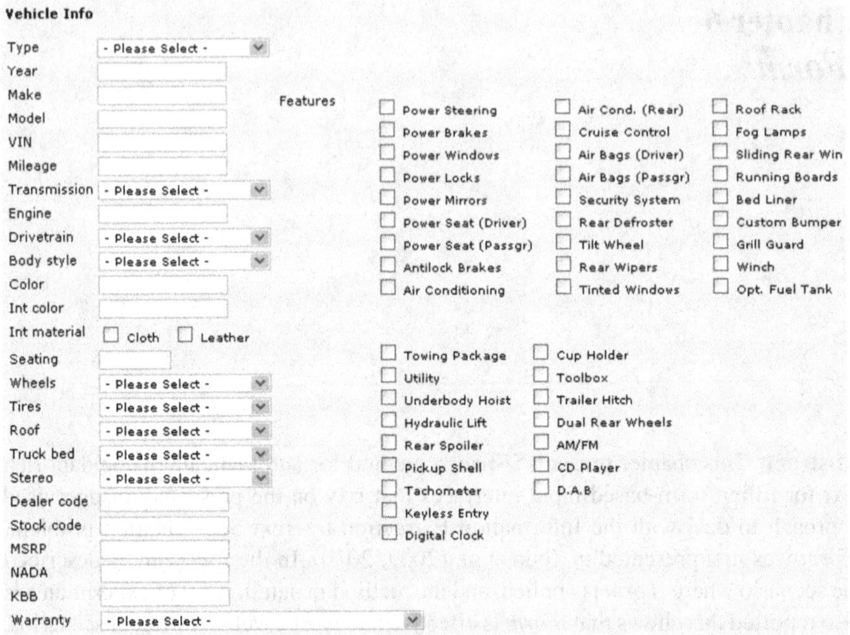

Vehicle Info

			Features			
Type	- Please Select - ⌄			☐ Power Steering	☐ Air Cond. (Rear)	☐ Roof Rack
Year				☐ Power Brakes	☐ Cruise Control	☐ Fog Lamps
Make				☐ Power Windows	☐ Air Bags (Driver)	☐ Sliding Rear Win
Model				☐ Power Locks	☐ Air Bags (Passgr)	☐ Running Boards
VIN				☐ Power Mirrors	☐ Security System	☐ Bed Liner
Mileage				☐ Power Seat (Driver)	☐ Rear Defroster	☐ Custom Bumper
Transmission	- Please Select - ⌄			☐ Power Seat (Passgr)	☐ Tilt Wheel	☐ Grill Guard
Engine				☐ Antilock Brakes	☐ Rear Wipers	☐ Winch
Drivetrain	- Please Select - ⌄			☐ Air Conditioning	☐ Tinted Windows	☐ Opt. Fuel Tank
Body style	- Please Select - ⌄					
Color						
Int color						
Int material	☐ Cloth ☐ Leather					
Seating				☐ Towing Package	☐ Cup Holder	
Wheels	- Please Select - ⌄			☐ Utility	☐ Toolbox	
Tires	- Please Select - ⌄			☐ Underbody Hoist	☐ Trailer Hitch	
Roof	- Please Select - ⌄			☐ Hydraulic Lift	☐ Dual Rear Wheels	
Truck bed	- Please Select - ⌄			☐ Rear Spoiler	☐ AM/FM	
Stereo	- Please Select - ⌄			☐ Pickup Shell	☐ CD Player	
Dealer code				☐ Tachometer	☐ D.A.B	
Stock code				☐ Keyless Entry		
MSRP				☐ Digital Clock		
NADA						
KBB						
Warranty	- Please Select - ⌄					

Fig. 6.1 A real Web form from the cars domain

2005 Honda new **Accord** Ex, Clean, very **low Mileage**, Maintained By Dealer! Vehicle Located in Stockton, Ca. Ad Id # 28147
This is a brand new car with **automatic transmission!**

Car with Air Conditioning, clock, **Cruise Control**, Digital Info Center, Dual Zone Climate Control, Heated Seats, Leather Steering Wheel, Memory Seat Position, Power Driver's Seat, **Power Steering, Power Breaks**, Power Passenger Seat, **Power Windows, Cup Holder, Rear Air Conditioning, Sunroof**, Tilt Steering Wheel, Original Owner, **Alloy Wheels**, Am/Fm, **Cd Changer**, Mp3, Satellite.

Contact Us At XXX-XXXX-XXXX more information Visiti xxx xxx motors

Fig. 6.2 An example of a car ad in free text

The method presented here receives a data-rich free text input (e.g., an offer or ad), such as the one illustrated in Fig. 6.2, and extracts implicit data values occurring in it that can be used to appropriately fill out the fields in a form-based interface. For practical purposes, the user could check if the fields were correctly filled by the system and make any necessary corrections before inserting the data into the underlying Web database.

Thus, the problem faced by *iForm* is automatically filling out the fields of a given form-based interface with values extracted from a data-rich free text document, or portions of such documents. In particular, it can be identified two subproblems: (a) extracting values from the input text and (b) filling out the fields of the target form using them.

Free text documents are treated as sequences of words $t_1, \ldots, t_N$, representing individual words or punctuation. The extraction task consists of identifying segments from the free text document, i.e., a sequence of contiguous words, which are suitable values for fields in the form. A segment s_{ij} is composed of words from $t_i, \ldots, t_j$, such that $i \le j, i \ge 1$ and $j \le N$. A valid set of values extracted from the input text must follow two conditions: (1) only a single segment can be assigned to each field in the form and (2) every extracted segment must be non-overlapping, i.e., there are no extracted segments s_{ab} and s_{cd} for $a < c$ such that $b \ge c$.

Most of the challenge of the form filling problem is related to subproblem (a), since suitable values are sparsely embedded in the text with other non-related strings. Furthermore, no particular format or order can be assumed for these values.

6.2 The *iForm* Method

iForm relies on the proposed approach to the IETS problem, presented in Chap. 3. In this case, the knowledge base is formed by previous values submitted to each form field. For simplicity, we refer to the user free text document or portions as *input text* from now on. Users may want to verify the form filled by *iForm*, make corrections, and then proceed with the request submission. After that, the new assigned values are stored in the knowledge base and considered when new input texts are provided by users.

The *iForm* method uses content-based features (Sect. 3.3) to extract text segments in the input text that are suitable for filling a given field in a form. *iForm* considers the following content-based features to perform the extraction task: (1) the Attribute Vocabulary feature described in Sect. 3.3.1, which exploits the common words often shared by values of textual form fields; (2) the Attribute Value feature that is similar to the Attribute Vocabulary feature, but instead of exploiting the common words, explores the common values often shared by values of textual form fields; and (3) the Attribute Value Format feature, described in Sect. 3.3.1, that exploits the common writing style often used to represent values of fields. Notice that for simplicity of notation, in this setting, the meaning of "Attribute" is similar to "Field."

Also, it is important to stress that these features mentioned above are computed based on the knowledge base generated with previous values submitted to the form, and also, no features from the input texts are considered. As shown by Toda et al. (2010), these features can be easily considered as probabilities in a probabilistic framework.

An interesting property regarding the *iForm* strategy is that it allows us to correctly identify segments in the input text that may not correspond to values previously entered in the field, as long as these segments include words typically found in the values of this field or have a format usually associated to the values previously used in that field.

Consider an input text I, which is composed of $N > 0$ words. Let S_{ab} be a segment, i.e., a sequence of words in I that includes words $t_a, t_{a+1}, \ldots, t_{b-1}, t_b$ $(0 < a \leq b \leq N)$. We consider S_{ab} as a suitable value for a field f if the score returned by the content-based features is above a threshold ε.[1] Considering L as the maximum segment length, there are $N * L - \sum_{i=1}^{L-1} i$ segments in a text with N words.[2] As latter detailed, $iForm$ deploys a dynamic programing strategy to avoid recomputing the scores for all pairs of segments and fields.

In the following it is shown how the content-based features can be used to extract text segments from input texts.

6.3 Using Content-Based Features

Given a text segment S_{ab}, $iForm$ decides if this segment is a suitable value of a given field of the form taking into account different content-based features g^k evaluated by feature functions of the form $g^k(S_{ab}, f)$. To combine these features, it is assumed that they represent the likelihood of the candidate value S_{ab} to occur as a value of the field f domain, according to the knowledge base. These content-based features are combined using a Bayesian disjunctive operator or, as described in Sect. 3.5.1.

Considering the content-based features described earlier, $iForm$ first computes the attribute vocabulary feature described in Sect. 3.3.1. The intuition for the usage of this feature is that the more concentrated the previous occurrences of a term are in a field, the higher the likelihood of this field being related to the term.

It can be noticed that the computation of the values of $g^k(S_{ab}, f)$ for every possible segment leads to a redundant computation which can be avoided by using dynamic programming. Based on this, it can be defined mp_{ij}, the matrix containing the features result of a field ft_k given segment s_{ij} as follows:

Let $mp_{ij} = P(ft_k|s_{ij})$, the following recurrence can be used to compute this feature:

$$mp_{ij} = \begin{cases} g^k(ft_k|s_{ij}) & i = j \\ mp_{i(j-1)} + mp_{jj} & i < j \end{cases} \tag{6.1}$$

The dynamic programing algorithm for solving this equation first computes the simplest case, that is, elements mp_{ij} such that $i = j$. The algorithm then computes elements in the first row from left to right, and proceeds to the following rows until all elements in the matrix are defined. This process is repeated for the matrices of each field.

---

[1] In all of the experiments, we performed a previous training and selected $\varepsilon = 0.2$.

[2] In the experiments L is no greater than 10.

The second feature considered by *iForm*, the attribute value feature, exploits common values often shared by form fields, instead of words. Its computation is similar to the computation of the attribute vocabulary feature, but, in this case, it considers the submitted values itself, and not the words that compose these values.

Finally, it can also be computed the value of $g^k(S_{ab}, f)$ considering the attribute value format feature. Notice that in this case, *iForm* computes how likely are the sequences of symbols representing the text segment S_{ab} to be a value of the field f.

During the experiments, it was verified that, in the Web form filling task, the attribute value format feature is less precise than the other content-based features. Indeed, the style information is helpful when token and value features fail to associate some segments to a given field. Because of this, we decided to use the writing style information as part of a refining process.

Thus, the mapping process, described in Sect. 6.4, uses the content-based features in two phases, and the attribute value format feature is not taken into account in the first phase. The first phase only combines the attribute vocabulary feature and the attribute value feature. In cases where the first phase fails on finding text segments to fields, *iForm* takes into account the attribute value format feature in the combination process (Sect. 3.5.1).

6.4 Mapping Segments to Fields

Let C_j be the set of segments S_{ab} such that $\ell(s, A)$ (see Eq. 3.10), which returns a score with the result of the combination of the content-based features, is above threshold ε. It can be said that C_j is a set of *candidate values* for field F_j.

The aim is to find a *mapping* $\mathcal{M}$ between candidates values and fields in the form-based interface with a maximum aggregate score, such that (1) only a single segment is assigned to each field and (2) the selected segments are non-overlapping, i.e., there are no segments S_{ab} and S_{cd} for $a < c$ in the mapping such that $b \geq c$. This is accomplished by means of a two-step procedure as follows.

In the first step, *iForm* begins by computing the candidate values for each field F_j, based only on the attribute vocabulary feature and the attribute value feature. Let $\mathcal{I}$ be a set composed of the union of the sets of candidate values C_j for all fields F_j. We refer to $\mathcal{I}$ as the *initial mapping*, which contains segment-field pairs $\langle S_{ab}, F_j \rangle$. It can be said that two pairs in $\mathcal{I}$ are in *conflict* if they violate any of the conditions above. Hence, the problem is finding a subset of value-field pairs in $\mathcal{I}$ without conflicts whose aggregate scores are maximum.

Finding the optimal solution for this problem requires assessing all possible subsets—an exponential number. In practice, *iForm* uses a simple greedy heuristic to find an approximate solution. First, *iForm* extracts the pair with the highest score from $\mathcal{I}$ and verify whether it presents conflict to any pair in $\mathcal{M}$ or not. If such a pair is non-conflicting, it is added into the final mapping. This process is repeated until every pair in $\mathcal{I}$ is extracted. This ends the first step.

In the second step, if any field remains not mapped to a segment, we use the attribute value format feature to try to find further assignments. *iForm* then repeats the mapping process, but now considering only pairs of segments and fields that were not mapped in the first step.

The two-step mapping was adopted after verifying through experiments that the attribute value format feature is less precise than the other two features adopted. On the other hand, it is still interesting to use writing style information when word and value features fail in associating some segment to a given field. Thus, we decided to use the style information as part of a refining process, which is performed in the second step of the mapping.

6.5 Filling Form-Based Interfaces

The last step in the *iForm* method consists of using the final mapping $\mathcal{M}$ to fill out the fields of the form-based interface.

In the case of text boxes, *iForm* simply enters each mapped text segment as a value into its corresponding field. For check boxes, *iForm* sets true for fields that were mapped in $\mathcal{M}$ and false for other check boxes. Since extracted values are rarely equal to items in pull-down lists, this type of field requires more work as we discuss in the following.

In the case of pull-down lists, *iForm* aims at finding an item such that its similarity with the extracted value is maximum.

iForm measures this similarity by using a "soft" version of the well-known cosine measure, named softTF-IDF (Cohen et al. 2003). Unlike the traditional cosine measure, softTF-IDF relaxes the requirement that terms must exactly match and yields better results in the problem. The softTF-IDF model also assesses the similarity between terms by using a similarity measure for strings s. In this way, given a value A and a pull-down list item B, we define $close(\theta, A, B)$ as the set of term pairs (a, b), where $a \in A$ and $b \in B$, and such that $s(a, b) > \theta$ and $b = \arg\min_{b' \in B} s(a, b')$; i.e., b is a term in B with the highest similarity to a.

The similarity between a value A and an item B in a pull-down list is defined as follows:

$$soft(A, B) = \frac{\sum\limits_{(a,b)\in close(\theta,A,B)} w(a, A) \cdot w(b, B) \cdot s(a, b)}{\sqrt{\sum\limits_{a\in A} w(a, A)^2} \cdot \sqrt{\sum\limits_{b\in B} w(b, B)^2}} \tag{6.2}$$

where $w(a, A)$ and $w(b, B)$ are the weights of terms a and b related to the value A and item B, respectively. $w(a, A)$ returns 1 if a occurs in A or 0, otherwise. For computing $w(b, B)$ we consider the inverse frequency of term b in the pull-down list, i.e, $N_L/\texttt{freq}(b, L)$, where N_L is the number of items in the pull-down list L and $\texttt{freq}(b, L)$ is the number of values in L containing term b.

6.6 Experiments

In this section is reported the results of experiments conducted with *iForm* on tasks of automatically filling form-based Web interfaces.

6.6.1 Setup

In all experiments performed here, a real form-based Web interface was simulated where each data-rich free text document is submitted at a time. Users manually verify its results and, if needed, correct minor errors. After that interaction, the submission will be completed and new added values will be considered when processing new submissions from this point on. Notice that the system was evaluated according to the errors produced on each iteration. In all cases there is no intersection between the sets of test submissions and the set of previously submitted documents.

Table 6.1 presents in detail each dataset used in our experimental evaluation. The column "Test Data" shows the number of input texts submitted to the form-based interface. The column "Previous Data" refers to the number of previous submissions that were performed prior to the test.

The *Jobs* dataset was obtained from RISE (Repository of online Information Sources used in information Extraction tasks).[3] The test set consists of 50 postings and the previous data consist of 100 postings previously annotated, as it is required for the experimental comparison with iCRF (Sect. 4.5.5). For the datasets *Cars* and *Cellphones* multi-field Web form interfaces and input data-rich text documents were both taken from *TodaOferta.com* auction website. Similar to *Cars* and *Cellphones*, for the *Books* datasets, the input data-rich text documents were taken from *TodaOferta.com*, but, in order to evaluate how good *iForm* adapts to form variations, the multi-field Web form interfaces were taken from distinct websites, *TodaOferta.com*, *IngentaConnect.com*, *Oupress.com* and *Netlibrary.com*, composing datasets *Books* 1 to 4, respectively. For the test in these experiments, real offers submitted to *TodaOferta.com*

Table 6.1 Features of each collection used in the experimental evaluation

Datasets	Test data	Previous data	Source—test data	Source—previous data
Jobs	50	100	RISE	RISE
Movies	50	10000	IMDb	Freebase and Wikipedia
Cars	50	10000	TodaOferta	TodaOferta
Cellphones	50	10000	TodaOferta	TodaOferta
Books 1 to 4	50	10000	Submarino	TodaOferta, IngentaConnect, Oupress, Netlibrary

---

[3] http://www.isi.edu/integration/RISE/

were used and automatically filled the corresponding form. In the case of *Movie Reviews*, a Web form was built and real short movie reviews collected from IMDb[4] (Previous Submissions), and from Wikipedia and Freebase (Test Submissions).

To evaluate the results of the experiments the well-known metrics precision, recall and f-measure, were used. These metrics were applied to evaluate the quality of filling a single field and a whole form submission.

In the case of text boxes, we calculate precision, recall, and f-measure at field level as follows. Let A_i be the set of all tokens (words) from the input text that *should* be used for filling a given field i in the form. Let S_i be the set of all tokens from the input text that were used for filling in this field i by the automatic filling method. We define precision (P_i), recall (R_i) and F-measure (F_i) as:

$$P_i = \frac{|B_i \cap S_i|}{|S_i|} \quad R_i = \frac{|B_i \cap S_i|}{|B_i|} \quad F_i = \frac{2(R_i.P_i)}{(R_i + P_i)}. \tag{6.3}$$

For pull-down lists, set A_i contains the item in the list of field i that *should* be chosen and set S_i contains the items that were chosen for field i. For check boxes, A_i contains the *correct* Boolean value for field i and S_i contains the Boolean value that was set for field i.

Submission-level precision (recall and f-measure), i.e., the quality of a whole submission, is calculated as the average of the values of each field used in this submission, observing that there are submissions in which not all fields are used.

Prior to all experiments, an evaluation of the sensibility of *iForm* was performed with respect to the threshold ε. Following, it was tested the *iForm* method with multi-typed Web forms for submissions of *Short Movie Reviews*, *Car offers*, *Cellphone offers* and *Book offers*. Next, it was evaluated, in turn, how the number and the coverage of the previous submissions impact on the performance of *iForm*.

Finally, experiments using a *Jobs postings* dataset were conducted for comparing *iForm* with a solution previously proposed to interactively filled forms (Kristjansson et al. 2004), which is referred to as *iCRF* in the experiments. *iCRF* is a method for interactive form filling based on CRF (Lafferty et al. 2001).

6.6.2 Varying ε

One important question in *JUDIE* is to determine the value of the threshold ε. Recall from Sect. 6.2 that *iForm* considers a segment S_{ab} as a suitable value for a field f if the score returned by the content-based features is above a threshold.

To study this parameter, we randomly selected 25 documents from each dataset and submitted them to *iForm* varying the parameter ε from 0.1 to 0.9. The results of the averaged submission-level f-measure achieved are shown in Fig. 6.3, where in

---

[4] http://www.imdb.com

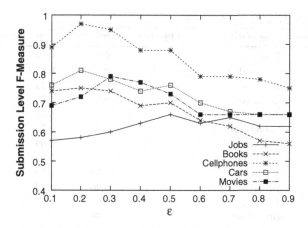

Fig. 6.3 Results obtained when varying ε

the case of *Books* datasets, the curve in the graph corresponds to the average results for *Books* 1 to 4.

It can be seen that the results may vary according to the domain, which suggests that a training adjusting step could be useful to produce optimized results. Notice however, that quite good results were achieved when using $\varepsilon = 0.2$ in the samples submitted. For the *Jobs* dataset, the best form-filling result was obtained with $\varepsilon = 0.5$. This can be explained by the small number of documents that compose the previous submissions, which requires a more restricted threshold. For all the experiments in this chapter, the value 0.2 for ε was used, including experiments with the *Jobs* dataset. We suggest the possibility of introducing a training step for future work.

6.6.3 Experiments with Multi-Typed Web Forms

To evaluate *iForm* within typical different form-based interfaces from distinct websites, we tested *iForm* with submissions from *Movie Reviews, Car offers, Cellphone offers* and *Book Offers*.

We grouped the results by the type of each field, i.e., text box, check box or drop-down list, according to their occurrence in each Web form. The results are presented in Table 6.2 by means of field-level and submission-level precision, recall, and F-measure.

As it can be noticed, *iForm* achieved high quality results in all datasets. In the case of car offers, as shown in Table 6.2 (Cars) the quality of the form filling task was almost the same for the text box fields and the check box fields.

Much better results were obtained for the case of cellphone offers, in which the F-measure average reached above 0.90, as shown in Table 6.2. As a consequence,

Table 6.2 Results for multi-typed Web forms

Domain	Type of field	# Fields	P	R	F
Movies	Text box	4	0.74	0.69	0.71
	Submission		**0.73**	**0.67**	**0.69**
Cars	Text box	5	0.78	0.73	0.76
	Check box	30	0.79	0.79	0.79
	Average		*0.79*	*0.78*	*0.79*
	Submission		0.77	0.73	0.75
Cellphones	Text box	2	0.89	0.69	0.78
	Check box	35	0.94	0.94	0.94
	Average		*0.94*	*0.93*	*0.93*
	Submission		**0.96**	**0.94**	**0.95**
Books 1	Text box	4	0.88	0.67	0.76
	Drop down	1	0.96	0.96	0.96
	Average		*0.90*	*0.73*	*0.80*
	Submission		**0.89**	**0.67**	**0.76**
Books 2	Text box	4	0.72	0.54	0.62
	Submission		**0.74**	**0.55**	**0.63**
Books 3	Text box	2	0.73	0.55	0.63
	Submission		**0.70**	**0.56**	**0.62**
Books 4	Text box	3	0.85	0.56	0.68
	Submission		**0.75**	**0.55**	**0.63**

submission-level f-measure result for this dataset was 0.95, which means that on average, more than 90 % of each submission was correctly entered in the Web form interface.

A detailed inspection on the offers entered by users in this interface revealed that the values available on these offers are usually more uniform than the values of car offers and movie reviews. This explains the excellent results obtained by *iForm* and corroborates our claims regarding the frequent reuse of data-rich texts for providing data to fill form-based interfaces on the Web.

In the case of the movie dataset the inspection of the text inputs entered revealed a large degree of ambiguity, since it is very common, for instance, to have actors that are directors and directors that are also actors. As well as this, movie titles contain ordinary words that appear within reviews not necessarily composing the title (e.g., "Bad Boys") and each review itself sometimes presents more than one movie title. In addition, names and titles that are entirely composed of terms not known from previous submissions frequently appear. In such cases, the style features play an important role. All these shortcomings make this dataset a real challenge. Similar difficulties were found in the *Books* datasets. Despite this, *iForm* presented good results. As shown in Table 6.2, precision levels are above 0.7 in all cases, and submission-level f-measure results for these datasets are above 0.6.

6.6.4 Number of Previous Submissions

In this experiment it can be verified how the performance of *iForm* behaves when the number of previous submissions varies. The result of this experiment is presented in Fig. 6.4, in which for each dataset, an increasing number of submissions, from 500 to 10000, was used and the average submission-level f-measure calculated resulting from running the form filling process over each collection.

As shown in Fig. 6.4, for the Movies and Books 1 datasets, the quality achieved by *iForm* increases proportionally with the number of previous submissions. The same behavior was observed for the other Books' datasets. Their results are presented in Fig. 6.5.

In the cases of the Cars and Cellphones datasets, it is important to notice that F-measure values stabilize at around 3000 previous submissions and remain the same until 10000 submissions. This shows that *iForm* does not require a large number of previous submissions to reach a good quality of results. Besides, even starting with a small number of submissions, *iForm* is able to help decrease the human effort in the form filling task. Notice that the expected volume of previous submissions in the application scenarios which motivated our work, i.e, sites such as *eBay* and

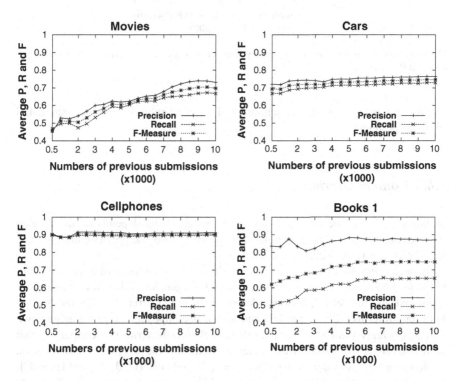

Fig. 6.4 Behavior of the form filling quality with the increasing of the previous submissions

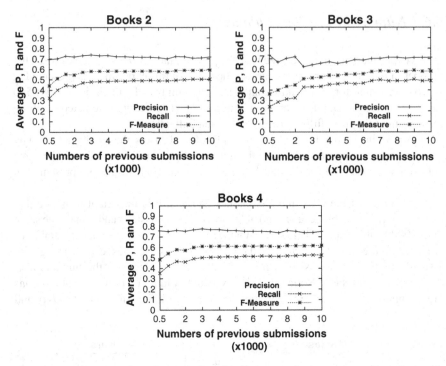

Fig. 6.5 Behavior of the form filling quality with the increase in the previous submissions with different Book forms

TodaOferta, is far higher than the number of previous submissions it was simulated in this experiment.

6.6.5 Content Overlap

In this experiment the aim is to study how much *iForm* depends on the overlap between the contents of the text inputs and the contents of the previous submissions, i.e., the known values submitted to each field.

In the *iForm* solution can be characterized three different forms of content overlap: (1) *Value Overlap*: the overlap between the set of complete values found in a given input text and the set of previously known values; (2) *Term Overlap*: the overlap between **all** terms on the input text and all terms composing the previously known values; (3) *Term-Value Overlap*: the overlap between the terms in the input text that compose values to be extracted and all terms composing the previously known values.

To exemplify, consider the input text $I = \{$"Brand New Honda Accord Hybrid"$\}$, from which the values "Honda" and "Accord Hybrid" are to be extracted for fields *Make* and *Model*, respectively. Suppose the following values are known for fields

Make and *Model*: *Make*= {"Honda", "Mercedes"} and *Model*= {"City","Civic Hybrid", "A310"}.

In this example, for input text I: (a) the value overlap is $1/2$, since from the two values to be extracted only one is known; (b) the term overlap is $2/5$, since from the five terms in the input text, only two are available in the known values; (c) the term-value overlap is $2/3$, since from the three terms composing values to be extracted from I, only two are previously known.

In Fig. 6.6 is presented the quality results of the experiment described in Sect. 6.6.3 for datasets *Movies*, *Cars* and *Cellphones* and *Books1*, showing different ranges of overlap, considering the three forms of overlap described above.

Figure 6.6j shows that for most of the inputs (36 out of 50) the value overlap is not greater than 50 %, and, despite that, the quality of the results in terms of precision, recall and f-measure is close to the quality obtained with a larger value overlap, 76–100 %, observed on three inputs. This is in accordance with the results presented in Fig. 6.6k, since most of the inputs have most of the terms present in the previous submissions.

Besides, the *Movies* datasets trends are similar to the ones in *Books1*. For the case of datasets *Cars* and *Cellphones*, notice that the term overlap is quite low in all input texts. This is due to a large number of useless terms typically available on such input text taken as whole. In these cases, however, useful terms appear within values to be extracted from these input texts, yielding to the good quality results achieved.

These results show an important property of the presented method: *iForm* does not rely on a high coverage of values in the previous submissions, as long as these submissions are representative of the domain.

6.6.6 Comparison with iCRF

Finally, we compare *iForm* and the interactive method proposed by Kristjansson et al. (2004), which we name here as *iCRF*, for the task of extracting segments from text inputs and filling a form. We took from the RISE Jobs collection a subset of 100 job postings already containing labels manually assigned to the segments to be extracted. These job postings form an adequate training set for *iCRF*, since this method requires examples of values to be extracted to appear within the context they occur. Thus, we could not use the remaining 450 job postings from the collection, for which extracted values are provided separately from the postings in which they occur. From the same set of 100 documents, we took the labeled segments to simulate submissions to the form-based interfaces for *iForm*. Notice that, unlike *iCRF*, *iForm* does not require the annotated input for training.

Next, we tested both methods using a distinct set of 50 documents, whose extraction outcome was available from RISE, allowing us to verify the results. These results are reported in Table 6.3 in terms of field-level F-measure.

For the experiment with *iCRF*, we used a publicly available implementation of CRF by Sunita Sarawagi and deployed the same features described in

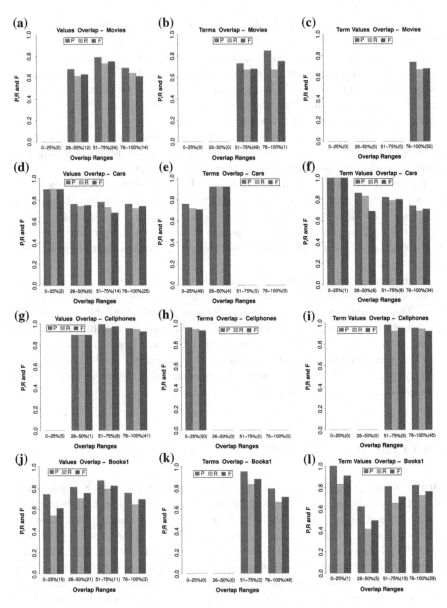

Fig. 6.6 Form filling quality on *Cars*, *Cellphones*, *Movies* and *Books1* datasets with different overlap ranges

Lafferty et al. (2001). Overall, these are standard features available on the public CRF implementation, e.g., dictionary features, word score functions, and transition features. Further, it was considered that the forms are filled in an interactive process,

Table 6.3 Field-level f-measure

Field	*iForm*	iCRF	Field	*iForm*	iCRF
Application	**0.82**	0.37	Platform	**0.47**	0.38
Area	0.18	**0.23**	Recruiter	**0.44**	0.22
City	**0.70**	0.65	Required degree	0.31	**0.59**
Company	**0.41**	0.17	Salary	0.22	**0.25**
Country	0.77	**0.87**	State	**0.85**	0.81
Desired Degree	**0.57**	0.37	Title	**0.72**	0.49
Language	**0.84**	0.69			

with the previously filled forms being corrected by a human and then being incorporated into the training set.

According to the results presented in Table 6.3, *iForm* had superior F-measure levels in nine fields, while *iCRF* had significant superior F-measure levels in four fields only, as indicated by boldface numbers. The lower quality obtained by *iCRF* is explained by the fact that segments to be extracted from typical free text inputs, such as jobs postings, may not appear in a regular context, which is an important requirement for CRF-based methods. For the case of *iForm*, this context is less important, since it relies on features related to the *fields* instead of relying on features from the *input texts*. In addition, *iForm* was designed to conveniently exploit these content-based features that from previous submissions that are related to fields. If we consider each submission as a whole, i.e., the submission-level quality, *iCRF* and *iForm* achieved, respectively, 0.46 and 0.59. Recall that, as we have seen, for one to apply *iCRF* to this problem, labor-intensive preparation of training data from a representative sample of text inputs is required.

References

Cohen, W., Ravikumar, P., & Fienberg, S. (2003). A comparison of string distance metrics for name-matching tasks. *Proceedings of the IIWeb Workshop on Information Integration on the Web* (pp. 73–78). Mexico: Acapulco.

Kristjansson, T., Culotta, A., Viola, P., & McCallum, A. (2004). Interactive information extraction with constrained conditional random fields. *Proceedings of the AAAI National Conference on Artificial Inteligence* (pp. 412–418). USA: San Jose.

Lafferty, J., McCallum, A., & Pereira, F. (2001). Conditional Random Fields: Probabilistic Models for Segmenting and Labeling Sequence Data. *Proceedings of the ICML International Conference on Machine Learning* (pp. 282–289). USA: Williamstown.

Toda, G., Cortez, E., da Silva, A. S., & de Moura, E. S. (2010). A probabilistic approach for automatically filling form-based web interfaces. *Proceedings of the VLDB Endowment, 4*(3), 151–160.

Toda, G., Cortez, E., Mesquita, F., da Silva, A., Moura, E., & Neubert, M. (2009). Automatically filling form-based web interfaces with free text inputs. *Proceedings of the WWW International World Wide Web Conferences* (pp. 1163–1164). Spain: Madrid.

Chapter 7
Conclusions and Future Work

Abstract This chapter presents the conclusions and discuss directions for future work based on the unsupervised approach presented here.

Keywords Information extraction · Unsupervised approach · Text segmentation · HTML · User feedback · Keyword-based queries

7.1 Conclusions

In this book, it was presented and evaluated an unsupervised approach for the problem of Information Extraction by Text Segmentation (IETS). This approach relies on knowledge bases to associate segments in the input string with attributes of a given domain by using a very effective set content-based features. The effectiveness of the content-based features is also exploited to directly learn from test data structure-based features, with no previous human-driven training, a feature unique to this approach.

It was studied different aspects regarding this approach and compared it with state-of-the-art IE methods. Results indicate that this approach performs quite well when compared with such methods, even without any user intervention.

Based on this approach, it produced a number of results to address the IETS problem in a unsupervised fashion. Particularly, it was developed, implemented, and evaluated distinct IETS methods. For the case where the input unstructured records are explicitly delimited in the input text, it was a method called On Demand Unsupervised Information Extraction (*ONDUX*) (Cortez and da Silva 2010; Cortez et al. 2010; Porto et al. 2011). Unlike previously proposed methods, *ONDUX* relies on a very effective set of content-based features to bootstrap the learning of structure-based features. More specifically, structure-based features are exploited to disambiguate the extraction of certain attributes through a reinforcement step. The reinforcement step relies on sequencing and positioning of attribute values directly learned *on-demand*

E. Cortez and A. S. da Silva, *Unsupervised Information Extraction by Text Segmentation*, 91
SpringerBriefs in Computer Science, DOI: 10.1007/978-3-319-02597-1_7,
© The Author(s) 2013

from test data. This assigns to *ONDUX* a high degree of flexibility and results in superior effectiveness.

This book also presents *JUDIE* a method for extracting semi-structured data records in the form of continuous text (e.g., bibliographic citations, postal addresses, classified ads, etc.) with no explicit delimiters between them. *JUDIE* is capable of detecting the structure of each individual record being extracted without any user assistance. This is accomplished by a novel Structure Discovery algorithm. It is also shown how to integrate this algorithm to the information extraction process by means of successive refinement steps that alternate information extraction and structure discovery. In comparison with other IETS methods, including *ONDUX*, *JUDIE* faces a task considerably harder, that is, extracting information while simultaneously uncovering the underlying structure of the implicit records containing it. In spite of that, it achieves results comparable to the state-of-the-art methods.

Finally, the unsupervised approach here presented is also exploited to create a method, called *iForm*, that is able to deal with the Web form filling problem. *iForm* (Toda et al. 2009, 2010) exploits values that were previously submitted to Web forms to learn content-based features. *iForm* aims at extracting segments from a data-rich text given as input and associating these segments with fields from a target Web form based on these features.

7.2 Future Work

The results achieved with the work here presented opens a number of possible paths for future development. Among them we may cite the following.

Generating Transductive Methods Using Domain Knowledge.

An issue we do not directly addressed in our work is how to better explore the presented unsupervised approach to create methods that are fully transductive, that is, that could learn content-based features from the input text in addition, or as an alternative, to the use of previously existing dataset. It would be interesting to investigate the possibility of generating sequence models specialized in a given input and to verify if these models converge to better extraction results.

Information Extraction from HTML pages.

Another interesting adaptation of the information extraction approach would be using it to extract information available in HTML pages. Although there are several alternative approaches to deal with this problem, they are generally too dependent on the regular use of HTML markup patterns. With the proliferation of alternative frameworks for content formatting such as the use of style sheets, scripting, and new languages such as HTML5, traditional extraction methods that rely on HTML markup can be severely affected. As the presented approach does not depend on particular markup features, we believe that it is possible to use it to not only to extract information but also to identify structured objects represented in HTML pages, such as product descriptions, recipes, etc.

Extraction Improvement Through User Feedback.

As many other approaches in the literature, the presented extraction approach is also subject to the occurrence of false positives (i.e., data wrongly extracted) and false negatives (i.e., data that should be extracted but that were not). We plan to incorporate some user feedback actions, hoping to improve the quality of the extracted data in cases where it is needed. For instance, we plan to use methods to identify possible extraction problems when the features we use do not reach a certain confidence level regarding the estimated quality of the extraction results. In these cases, the user could be required to provide high confident information that can be used as a feedback for the improvement of the process.

Structuring Keyword-Based Queries.

A typical application of information extraction methods is structuring queries expressed as a sequence of keywords, which are common in search engines. The main goal here is to correctly assign attribute names to terms provided in a keyword-based query (Li et al. 2009; Mesquita et al. 2007). We believe that it is possible to use the presented approach to address such a problem in an unsupervised fashion, i.e., with no user intervention. In fact, this work is currently been carried out, focusing on the problem of product search.

References

Cortez, E., da Silva, A., Gonçalves, M., & de Moura, E. (2010). ONDUX: On-demand unsupervised learning for information extraction. In *Proceedings of the ACM SIGMOD International Conference on Management of Data Conference, USA* (pp. 807–818). Indianapolis, USA.

Cortez, E., & da Silva, A. S. (2010). Unsupervised strategies for information extraction by text segmentation. In *Proceedings of the SIGMOD PhD Workshop on Innovative Database Research, USA* (pp. 49–54). Indianapolis, USA.

Li, X., Wang, Y.-Y., & Acero, A. (2009). Extracting structured information from user queries with semi-supervised conditional random fields. In *Proceedings of the International ACM SIGIR Conference on Research & Development of Information Retrieval, USA* (pp. 572–579). Boston, USA.

Mesquita, F., da Silva, A., de Moura, E., Calado, P., & Laender, A. (2007). LABRADOR: Efficiently publishing relational databases on the web by using keyword-based query interfaces. *Information Processing and Management, 43*(4), 983–1004.

Porto, A., Cortez, E., da Silva, A. S., & de Moura, E. S. (2011). Unsupervised information extraction with the ondux tool. In *Simpsio Brasileiro de Banco de Dados, Brasil*. Florianpolis, Brasil.

Toda, G., Cortez, E., da Silva, A. S., & de Moura, E. S. (2010). A probabilistic approach for automatically filling form-based web interfaces. *Proceedings of the VLDB Endowment, 4*(3), 151–160.

Toda, G., Cortez, E., Mesquita, F., da Silva, A., Moura, E., & Neubert, M. (2009). Automatically filling form-based web interfaces with free text inputs. In *Proceedings of the WWW International World Wide Web Conferences, Spain* (pp. 1163–1164). Madrid, Spain.